'This publication includes everything you need to know about these remarkable villages and their global history, from pirates to poets.'

Dr Alan Sandry, Swansea University

BUCKED IN THE YARN

THE UNIQUE HERITAGE OF COKER CANVAS

Bucked in the Yarn – The Unique Heritage of Coker Canvas.
Published in Great Britain in 2024 by Graffeg Limited.

Text by Terry Stevens copyright © 2024.
Produced by Graffeg Limited copyright © 2024.

Graffeg Limited, 24 Stradey Park Business Centre, Mwrwg Road, Llangennech, Llanelli, Carmarthenshire, SA14 8YP, Wales, UK.
www.graffeg.com.

Terry Stevens is hereby identified as the author of this work in accordance with section 77 of the Copyright, Designs and Patents Act 1988.

A CIP Catalogue record for this book is available from the British Library.

All rights reserved. No part of this publication may be reproduced, stored in a retrieval system or transmitted, in any form or by any means, electronic, mechanical, photocopying, recording or otherwise, without the prior permission of the publishers.

ISBN 9781802586978

1 2 3 4 5 6 7 8 9

GRAFFEG

BUCKED IN THE YARN

THE UNIQUE HERITAGE OF COKER CANVAS

TERRY STEVENS

Dedication

In admiration of the endeavour and resilience of the past generations of villagers who toiled to create a heritage that deserves wider recognition.

In memory of my parents, Molly and Eric Stevens, for their love and for the wonderful childhood they gave me in East Coker.

'In my beginning is my end, in my end is my beginning'.
T. S. Eliot, *East Coker*

Above: Eric and Molly Stevens.

Above: Sample of No. 1 Coker Canvas, 1812.

Acknowledgements

This book is testimony to the endeavours of so many villagers whose hard work and ingenuity over the centuries is the raw material for this story. It is, however, the diligence and commitment of local historians and others in the community that has ensured the history and heritage of these villages has been archived, recorded and conserved. At the heart of this has been the extraordinary efforts of the team behind the successful restoration of the Dawe's Twineworks and the activities of the Coker Rope and Sail charities, led by the indefatigable Ross Aitken. I am indebted to Richard Sims, who is firmly established as a leading historian of all aspects of the flax and hemp textiles of the region. His work is one of the main sources for much of this book and his editorial guidance has been extremely generous. There are others whose keenness to share the story of East, North and West Coker over the years has created a library of tales and pictures that pepper these pages – David Shorey, Nadine and Michael Dodge, Beatrice Hackwell, Abigail Shepherd, John Snelling, Chris Barker, Sir Mathew Nathan and, of course, David Foot (the award-winning sports journalist and son of East Coker). I am especially indebted to Bob Osborn, the wonderful historian who curates the Yeovil A-Z Virtual Museum and the anonymous author of the East Coker neighbourhood plan.

To all who have granted permissions and supplied photographs, please accept my sincere thanks.

Diolch yn fawr iawn – thank you – to the team at Graffeg, especially Joana Rodrigues and Daniel Williams, and to its founder, Peter Gill, thank you for believing in this project from the start and driving me to make it happen.

Finally, *diolch yn fawr iawn* to Catrin, Mari and Non, Ifan, Mike and the grandsons (Jac, Elis, Gruff, Owain *a* Moi) for embracing the love of East Coker.

Contents

Foreword by Jim Hartley, CEO, Ratsey & Lapthorn	8
A Message from Ross Aitken, Founder of the Coker Rope and Sail Charities	10
The Essence of the Cokers	14
Introduction by Terry Stevens	16

Flax and Hemp – The Essential Raw Materials 38

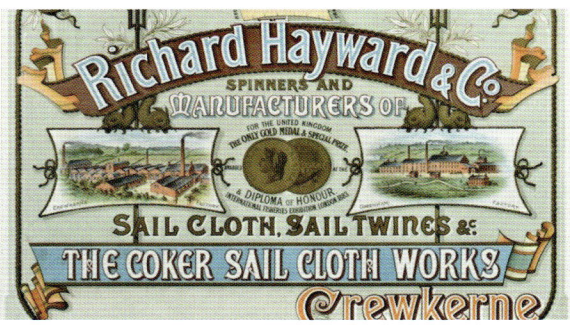

Bucked in the Yarn, Not in the Piece 20

The Making of a Global Brand 48

The Place 24

Giving the Past a Future and the Future a Past: Dawe's Twineworks 68

The Sailmakers, HMS *Victory*   78
and the America's Cup

Glossary of Terms – A Language of   112
its Own

The Pirate and the Poet   92

The Sailmaker's Tools   113

East Coker – *Esse Quam Videri*   114

Bibliography, Sources of Information   116
and Further Reading

Photo Credits   117

Grants, Supporters and Sponsors   120

About the Author   128

Musings and Mysteries   108

Foreword

Bucked in the Yarn is a meticulously woven tapestry of history that traces the remarkable journey of the villages of East, North and West Coker and their renowned contribution to the canvas industry. As the custodian of the world's oldest sailmakers, Ratsey and Lapthorn, it is an honour to lend my voice to this wonderful narrative, one that holds a special significance not just to ourselves but also in the annals of this country's maritime heritage.

For centuries, Coker has stood as a beacon of craftsmanship, its legacy intricately entwined with the production of Coker Canvas – a fabric renowned for its durability, strength and unparalleled quality. This book, however, delves far deeper into the heart of the villages and its people, unravelling the threads of time to reveal the stories of the skilled artisans who meticulously crafted each yard of canvas, imbuing it with a legacy that transcends generations.

At Ratsey and Lapthorn, we have had the privilege of harnessing the unmatched Coker Canvas for our sails, whether for the sleek yachts competing in the storied races of the America's Cups or for the warships of the Royal Navy endeavouring to 'rule the waves' in the name of king and country. Our sailmaking heritage and East Coker's history are inseparable, a testament to the enduring craftsmanship and timeless artistry that defines our shared legacy.

Within the pages of *Bucked in the Yarn*, you will embark on a journey through time, where the echoes of the past reverberate through the cobblestone streets and the looms hum with the rhythmic cadence of creation. It is a tale of resilience, innovation and the unwavering spirit of a community bound by its commitment to excellence.

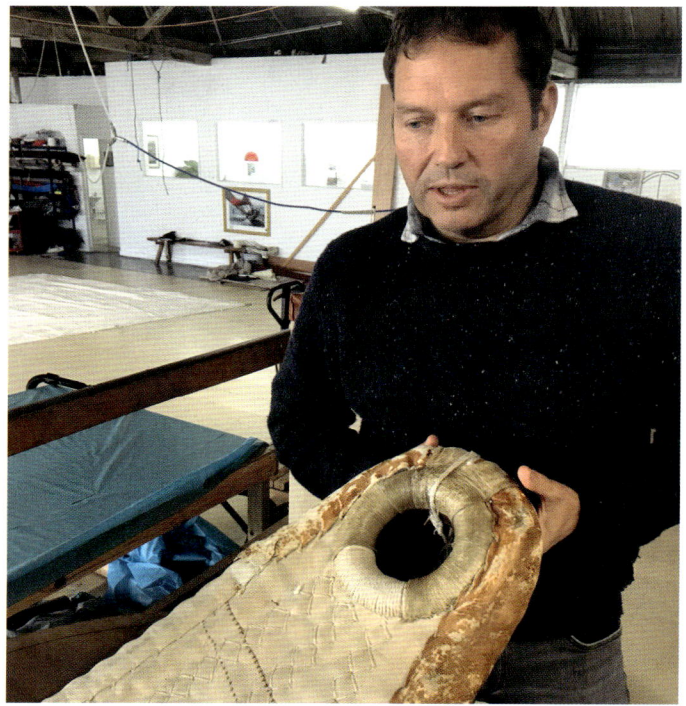

As you read, you will come to appreciate the indelible mark that Coker has left on the world. *Bucked in the Yarn* is not just a chronicle of these three villages; it is a testament to the power of craftsmanship and the timeless allure of a bygone era.

Fair winds and following seas,

Jim Hartley
CEO, Ratsey & Lapthorn

'There is only one standard of work in this loft, that is the very best.'
TOM W. RATSEY, 1833

A Message from Ross Aitken, Founder of the Coker Rope and Sail Charities

How a plan to save the Dawe's Twineworks became a means of telling the extraordinary story of Coker Canvas.

Almost 20 years ago, a small group of local volunteers came together to save the Dawe's Twineworks in West Coker. It soon became apparent to us that this restoration project was about more than saving an extraordinary heritage building at risk. The Dawe's Twineworks was the emblem of a much bigger story of the historical, social, commercial, technological and cultural context of the villages of East, West and North Coker.

We needed a vehicle to advance the celebration of our unique shared heritage. A means to enhance the public and especially our own communities, awareness of the need to preserve and conserve the twineworks but also the imperative to collect, research, educate and interpret the story of how local soils were especially good for growing flax and hemp and how this was used in the rope, sailcloth and netting industry of the three villages and how Coker Canvas became a global brand.

In July 2007 the Coker Rope and Sail Trust (Charity Number 1120031) was established to deliver these lofty ambitions. Seven years later, the Coker Rope and Sail Community Interest Organisation (Charity Number 1172029) was created to deliver the restoration of the twineworks. An extraordinary voluntary effort overseen by the two sets of trustees and supported by the Headley Trust, the National Heritage Lottery Fund, The Prince's Trust, the Andrew Lloyd Webber Foundation, South Somerset Council and Somerset County Council have successfully achieved the restoration of the Dawe's Twineworks.

It is now open to the public. It has a community café. It hosts numerous educational and special interest groups and The Od Arts Festival along with many community events. My sincere thanks to all that have been and are still involved in this ongoing project – especially my fellow Trustees.

There is much still to be done, and this book is an important step to help us spread the word about this important heritage story.

There is one person who has inspired me and two long-standing village friends to drive this heritage story forward. That person was Hugh Prudden.

Hugh Prudden was an exceptionally kind, charismatic and dedicated geography teacher at the then Yeovil Grammar School. A number of years ago I financed a booklet of his on the geology of the main buildings in Yeovil and while he led a group of local dignitaries, headed up by the Mayor, he noticed I was not paying attention, so, as a typical schoolmaster, he asked me to the front to give a detailed account of the geology of the walls of Barclays Bank.

In one of his classes, he had two students who have had quite different and interesting, careers and are still great friends. Hugh inspired them both. Terry Stevens, from East Coker, with a love of travel and an interest in landscape that was fuelled by Hugh Prudden, founded his multi-award-winning international tourism consultancy. Always a loyal friend, his strong determination can be shown in his lifelong support for Yeovil Town Football club – the Glovers.

The second student is Roger Bastable, from West Coker, who made a million, lost a million, then made another and today runs a successful hotel and restaurant. Still great friends, they typify the spirit of East and West Coker: going their own thing but working together.

Over recent years, Roger has been a loyal supporter of the work of the Charities whilst Terry has helped us in so many ways to develop Dawe's Twineworks in West Coker.

The importance of this book is that it puts into context the great importance of Coker Canvas – a global brand spawned from these small villages in South Somerset, some 20 miles from the sea – villages that saw great scandals and are the focus of a story full of intrigue, controversy and international impact.

Ross Aitken
Chairman, Coker Rope and Sail charities, West Coker

Left: Ross Aitken.
Top: Sign at North Coker.
Bottom: Roger Bastable, Haselbury Mill.

Above: Map of the villages, Ordnance Survey sheet 177, Taunton & Lyme Regis, 1930.

Right: Dawe's Twineworks (restored), West Coker.
Below: Drake's factory, East Coker, c. 1950.

The Essence of the Cokers

As we have come to expect from his previous work, Terry Stevens skilfully captures the essence of the Cokers. There is warmth and wisdom in his writing. He draws upon his deep, intimate knowledge of the place of his birth and successfully merges this with numerous and varied historical sources. The story is told in sufficient detail, with a comprehensible tone that will be accessible to specialist and lay readers alike. This is a much-needed volume that tells a hitherto sidelined story. This publication includes everything you need to know about these remarkable villages and their global history, from pirates to poets.

We are introduced to a plethora of fascinating, remarkable and ground-breaking characters. The Cokers are the fulcrum on which its notable inhabitants reached out to the world and to where many visitors arrived.

Dr Alan Sandry
Senior Lecturer in Identities and Political Philosophy, School of Management, Swansea University.

'a much-needed volume that tells a hitherto sidelined story'

Cider making at Isles Farm, 1966 (Bob Richards, George Ford and Bill Richards).

An Absorbing Human-interest Story

Reading this book is a revelation. It tells the fascinating story of East Coker, North Coker and West Coker, three small Somerset villages that achieved global importance for their production of sailcloth of exceptional quality using locally grown flax and hemps. Made in the area from the Middle Ages onwards and reaching peak production in the late 19th century, 'Coker Canvas' won an enviable international reputation for being the best sailcloth in the world. It was never the cheapest but was demonstrably stronger and more durable than its rivals. Its superiority led to its being adopted throughout the world by navies and merchant fleets, including those of Britain, luxury yachts and, not least, competitors in the America's Cup. The rich history and heritage of the Coker villages is little known, and its significance deserves greater recognition and appreciation. It is a story that is well worth telling and its resonances will undoubtedly be of interest to a wider audience.

This engaging and often thought-provoking volume tells this story very well, written as it is in a lively, accessible style and enlivened by flashes of humour and quotations from contemporary sources, poetry and the memories of ordinary Coker people. *Bucked in the Yarn* is an absorbing human-interest story, a tale of innovation, enterprise, fallibility, intrigue, controversy and endurance, and of local pride and determination that among many other things led to successful efforts to record and preserve the history and create a heritage centre. The book shows vividly how the wider world impacted on the Coker villages and their inhabitants – in the form of wars and technological changes, for example – and how, at the same time, this locality and its people made an impact on the wider world. It also conveys evocatively a strong sense of place of the three villages and explores absorbingly and enthusiastically the interplay between the local, the global and the individual, notably among the latter the explorer (among many other things) William Dampier and the poet T. S. Eliot.

I hope that a wider audience can learn about and appreciate the remarkable global heritage of these three small but extraordinary Somerset villages.

Bill Jones
Emeritus Professor of History,
School of History, Archaeology and Religion, Cardiff University.

'A tale of innovation and enterprise'

Introduction

East Coker Primary School, where I spent the years 1957–1963, dates from 1851. To the north, classrooms overlooked the birthplace of William Dampier, one of the greatest seafarers in British history, and, close by, the home of Edward Taylor, one of the great sailcloth manufacturers in the village.

To the south of the school was Taylor's former sailcloth factory, twineworks and ropewalks. On the school's western flank stood a row of canvas weavers' cottages and to the south and east was Coker Water – the eponymous crooked stream that gave the villages their name.

For a few years, my mother worked in the office of Drake's webbing factory, the successor to Taylor's factory. During my secondary school holidays in the 1960s I was employed to clean the factory's looms – a filthy, poorly paid job. My dad, following in the footsteps of his father and uncle, became the village carpenter. Their workshop was in a former yarn barton surrounded by cider apple trees. As a carpenter, Dad was, by default, the village undertaker. In 1965, this meant that he was invited to undertake the responsibility for the interment of the ashes of T. S. Eliot, the Nobel Prize-winning poet, whose ancestors were from East Coker.

Throughout my time at East Coker School, no one told us about the extraordinary people with connections to our village. We were never taught about the remarkable history associated with making twine, rope and sailcloth that was all around us. Familiarity does breed invisibility. Too many things just taken for granted maybe?

All this time, I was surrounded by intriguing place names (Bubs Pool, Pittes, Pavyotts, Burton Barton), passing buildings with unusual titles (the bucking house, the barton and the ropewalk), wondering about the meaning of imponderable words (retting, bolling and scutching), playing football in fields named the South Seas and Guiana and living next door to families (the Helyars, the Drakes, the Rendalls, the Goulds, Dawes and the Maudslays) whose forebears had been instrumental in shaping this important story in the history of Britain and beyond.

It was not until 1974, having been away from the village for six years, that the importance of this historical legacy dawned on me and, together with a small group of friends, we started the East Coker Society to raise awareness of this significant heritage. Fifty years on, it is my privilege to play a small part in the restoration and conservation of

Dawe's Twineworks – a surviving memorial of the heritage of sailcloth and rope making and the only fully working Victorian twine factory in England – by the Coker Rope and Sail charities. This project has been driven by the visionary leadership of its founder and chair, Ross Aitken, the son of my former primary school head in the 1960s.

Both my mother and father served the war years in the Navy. Despite the emergence of holidays for all at home and overseas in the 1950s, they showed no desire to travel. They were content with their square mile. One generation on, I have been fortunate to have worked on tourism projects in over 50 countries around the world. This interest in travel was inspired by the writings of William Dampier, the explorer, buccaneer and hydrographer who was born in East Coker in 1651, who wrote that 'A lack of prejudice and an inextinguishable curiosity makes one an instinctive traveller', as well as by T. S. Eliot's statement, which to me sounds like the perfect definition of tourism, 'for the Odyssey of the human spirit ... men require of their neighbours something sufficiently akin to be understood, something sufficiently different to provoke attention, and something great enough to command admiration.' This text appears in *Towards the Definition of Culture* (Eliot, 1948). It is also Eliot who sets the challenge in his poem *East Coker*, so relevant to me now, that 'Old men ought to be explorers, here or there does not matter. We must be still and still moving.'

Terry Stevens

Above: Drake's factory, c. 1963.
Left: The River Od/Coker Water – the crooked stream near Ford Flour Mill (sometimes known as Lewis' North Coker Mill).
Page 18-19: East Coker from St Michaels Church.

Bucked in the Yarn, Not in the Piece

For over 300 years, Coker Canvas was the global standard for sailcloth, adopted by the British Navy and countries around the world.

This international status was achieved because of its inherent strength and durability, born out of a unique process of soaking, or bucking, the yarn in an alkaline lye rather than bucking the whole piece of woven canvas. This process was key to the remarkable success and superiority of Coker Canvas, giving rise to the local phrase that defined it: 'Bucked in the yarn, not in the piece'.

Bucking took place in the bucking house, a single-storey building close to running water that contained furnaces and cisterns for boiling the yarn. The yarn was initially steeped in hot water, then bucked in the alkaline lye before being dried on grass in bleaching fields, or on wooden rails in yarn bartons, for two to three weeks. The process was then repeated four to five times before the yarn was finally soured with milk for another three weeks to neutralise it. It was this method of sailcloth production that was patented by the villagers of the Cokers.

The long history of making Coker Canvas from flax and hemp fibres was mirrored in the success of producing twine and rope in these villages. The quality and durability of these products created an enviable international reputation that meant that the products, especially the canvas, were in demand around the world. The finished bolts of canvas and yarn were exported to America and Europe and London-based merchants imported bales of flax from the Baltic region for use by local entrepreneurs. As a result, many lead seals from these bales have been found in the villages in recent years. This international buying and selling of goods brought traders from far and wide to the Cokers. Perhaps the voices of merchants from Arkhangelsk (Russia) and Riga (today the capital of Latvia) would have been heard on the village streets, their currencies exchanged in local hostelries for refreshment and other hospitality 'services'. It seems possible there was a red light area in West Coker!

West Coker is the Narrowbourne of Thomas Hardy's *A Tragedy of Two Ambitions,* published in 1894, while East Coker was rhapsodised by J. B. Priestley for its charms and is the ancestral home of two giants of culture and history: T. S. Eliot (1888–1965), one of the most important poets of the 20th century, and William Dampier (1651–1715), the explorer, pirate, navigator and naturalist, the first Englishman to set foot on what is today Australia and the first person to circumnavigate the world three times.

Above: The Coker Sail Cloth Works poster.
Right: Lead seal in Cyrillic script used on bale of semi-processed hemp from Baltic Russia.

It would be nice to think that Eliot's ancestors may have sailed to the Americas on ships using Coker Canvas, or that Dampier's dramatic exploits were on vessels suited with sails made from sailcloth from his home village. Whilst we can romanticise that notion, we can be certain that Lord Nelson, the German Kaiser Wilhelm II and Sir Thomas Lipton made certain that their commands sailed under Coker Canvas.

Despite this remarkable history, Lt. Col. Sir Matthew Nathan, in his formidable work *The Annals of West Coker* (1930), made the astonishing statement, 'Nothing much ever happened in this part of England where the Coker villages lie. No great man was ever born or died here. No battle was fought near it nor did any constitutional crisis have its rise in the neighbourhood. It was never the centre of great industry nor the source of wide-spreading trade. No relic of saint nor monument of art nor scene of natural beauty ever attracted visitors to it.'

How wrong could such an esteemed historian be? Was this irony or a case of mistaken identity? Was it purposeful modesty or a genuine lack of appreciation of the significance of what happened here in this corner of England, the Wessex of the Saxon kings and Thomas Hardy?

This book challenges Nathan's opinion, putting on record the remarkable and unheralded story of the villages that gave us Dampier and Eliot and a reputation built on the craft of artisans. It is the essential story of ordinary people – Job Gould, Israel Rendall, William Dawe, Edward Taylor, Constance Pley and many others – achieving the extraordinary in three small south Somerset villages, differentiated by the points of the compass but sharing the same familial name – Coker. Individual and collective endeavour, enterprise and innovation sustained over many centuries ensured that these villages were the crucible for the creation of a global brand based on an international reputation for producing the best sailcloth in the world.

Coker Country

In 1969, Bryan Little, writing in *Portrait of Somerset*, refers to 'Coker Country', ranking its importance in the development of sailcloth as being equivalent to the impact of the engineering and shipbuilding industries on the River Clyde in the late 19th century. This being the case, then Coker's contribution to the Royal Navy's global dominance on the high seas and the success of global trade and exploration during the 17th and 18th centuries must be regarded as incredibly significant.

So, how did this cluster of highly specialised maritime-related and influenced activities come to be so densely developed and remain in existence for such a sustained period of time in a small corner of south Somerset, an area some 20 miles from the nearest coast?

The base conditions were right. The underlying geology produced soils perfect for growing flax and hemp, whilst numerous springs fed small, gently flowing brooks and streams essential for processing the fibres of these plants. The lords of the manors encouraged their tenants to manage their plots productively. Entrepreneurship and innovation were nurtured with local finances and the enterprise supported by high-level political and naval influencers.

In addition, there had always been strong, well-established connections with the sea and maritime heritage. The maritime connections are deeply embedded and sometimes quite unusual. For example, in the north-western part of the parish of East Coker, within the Naish estate, part of the village that William Dampier would have known and visited, are three fields whose names are curious. They are known locally as the 'maritime fields'. They are Culliver's Grave, Doggen Sheet and Guiana, and can be found on the way through the sandy lanes towards West Coker. Guiana was the name of the exploitative colonial enterprise of Sir Walter Raleigh, of Sherborne

Above: St Martin of Tours' Parish Church, West Coker.
Right: The Square, West Coker, showing the Cave's store and post office, c. 1900.

Lodge and Castle, in the 1590s, according to one source. Culliver's Grave could refer to a wandering spirit of one John Culliver, who was lost at sea in a vessel named *Course* in 1658. A doggen sheet may have been a type of canvas, or part of a sail, and further fields between here and West Coker were known as South Sea and the Great South Sea.

A special place with an incredible heritage

Sixty years ago, in a promotional guidebook published by the Great Western Railway, this area was described as 'a romantic and historic corner of ideal England: a land of soft hills, peace, stillness, blossom in springtime and deep thatch'. The guide began by stating 'This is Home' and went on to eulogise that 'here one finds an oldness, a kindness, and a wisdom: things in part of the countryside and the dwellers within it. A place full of fragrant legends – from King Arthur to the Odcombe "leg stretcher".'

This is the story of this place, the unique products and the people that made this incredible heritage. It is a story about the way the villagers interacted with the landscape and nature, experimented with new ideas and embraced industrial innovation. Their progress was determined by national and international politics, global trade and technical advancements beyond their control. Disreputable business practices, dubious patronage and industrial espionage attempted, but did not succeed, in derailing the determination of the local enterprising spirit.

The Place

As villagers, we often reflected that it would be pleasing to record a full compass set of Cokers. There are East, West and a North Coker but no South Coker – at least not today. What is in a name? Over time the name Coker has appeared in official records and on various maps as Cochra, Cocre, Couker, Caucor, Cokr, Kokre and Cocker – indeed, 'my cocker' is still used as a term of endearment for someone from the area.

Etymologists offer varying origins of the place name. In the Anglo-Saxon period, names of local features, Saxon personal names or sometimes both gave rise to the names of settlements. For some, Coker originates from 'cochre' due to the red colouring of local soils – a theory supported with an area known as 'Redlands'. The more compelling story is that Coker is derived from the Old English, 'crocian' or a 'crooked stream' – the Coker Water.

Issuing from a spring near the village of Odcombe, the Coker Water is now more commonly known as the River Od. The Od is modest in both its form and power – hardly worthy of being called a river. It is never in a hurry to get anywhere as it meanders in an unruly fashion through the shallow valley of the Cokers. It is often hidden from view amongst tree-lined banks of willow, beech and ash, wending its way through withy beds and moors before joining the River Yeo. Along its journey it is fuelled by numerous natural springs (including the Holywell, the Blackwell, the Beauty and Peter's Hole), whilst its leets diverted its waters to drive numerous corn and flax mills and created ponds for retting and bucking the flax. The river is the consistent thread weaving these communities together. Be it the soil or the crooked flow of the Coker Water, both were to play a key role in the story of Coker Canvas.

Timeless villages, ancient landscape

The triptych of villages always shared a deeply entwined history. It is a story that reflects the way the fibres from the locally grown flax were twisted to create a much stronger yarn to make the twine, cordage and ropes. Whilst intense rivalries prevailed in local courting customs and amongst the villages' football, cricket, skittles and darts teams, intense cooperation fuelled economic success. Together, the villagers built their hard-earned, enviable international reputation, which was nurtured and maintained for over 300 years, despite numerous attempts by politicians, governments, unscrupulous entrepreneurs and other centres of industry to usurp their supremacy in the discipline of making quality canvas and twine. Eventually, however, new technologies, alternative new materials and unforeseen twists of fate brought an end to the Coker dominance in the market for sailcloth.

Although much has altered, little has changed

The Cokers are three neat, pretty, unassuming villages that embrace their sibling hamlets. They are set in an ancient landscape once claimed by the Romans, quintessentially English and undoubtedly Somerset. Today, they still feel like villages with timeless characteristics. As T. S. Eliot observed, 'In succession houses rise and fall, crumble, are extended, Are removed, destroyed, restored'. Surviving between the cottages are remnants of ancient apple orchards with their wonderfully named traditional species, such as the Bloody Butchers, the Yellow Horner and the Coker Seedling.

Grand picturesque manor houses, the square bell-towered churches of St. Martin of Tours' in West Coker and

Left: East Coker cricket team at the Recreation Ground, c. 1920.
Above: The Helyar Arms, East Coker.

St Michael and All Angels' in East Coker (today celebrating over 700 years as a site of continuous worship) and their attendant vicarages, a myriad of small non-conformist chapels, the substantial homes of the yeoman farmers, small-thatched weavers' cottages, almshouses and four-storey water mills exist cheek-by-jowl. Those of age are built of what many regard as the loveliest stone in England – Ham stone – so beautifully described by an observer as 'the colour of biscuit sprinkled with gold and able to trap sunlight in day and release it at dusk'. This is a sandy limestone composed of broken seashells that is quarried from nearby Ham Hill.

They are presented as picture postcard cottages with paths lined with delphiniums and hollyhocks and rambling roses in the front gardens. Most retain their pragmatic vegetable plots born of centuries of self-sufficiency, complete with a pigsty and a chicken coup. Their charm and beauty witnessed by the tourist's gaze today hides the reality that, even as late as 1958, many of the cottages were without an indoor toilet, some still used oil lamps and drinking water was often drawn from a well or a communal tap – as was the case outside our family cottage.

Iron age hill forts and Vespasian villas

There were villages of a kind here in the Bronze Age as the development of agriculture demanded settled communities. Iron Age hill forts were established at Ham Hill and at nearby Cadbury Castle (a site often associated with being King Arthur's Court of Camelot and an inspiration for John Steinbeck's interest in the Arthurian legends). However, little is known of life in the area before the Romans. The remarkable gold Yeovil Torc, clearly the work of a skilled goldsmith between 1300 BC and 1100 BC (now in the Museum of Somerset, Taunton), was discovered in soil by Henry Coles in 1909 in a house on the northern boundary of the parish of East Coker.

Somerset – the land of the summer people – was invaded from the south-east by Legio II Augusta (the Second Augusta Legion) – a name kept alive until recently with

the AgustaWestland Anglo-Italian helicopter company in Yeovil – under the command of the future emperor Vespasian. His legion captured the hillforts of the Durotriges at Ham Hill and Cadbury Castle. A Roman fort was set up at Ilchester (Lindinis) as the Romans established a defensive boundary along the new military road known as the Fosse Way, which ran through Bath, Shepton Mallet, Ilchester and south-west towards

Above: Roman mosaic recovered from the site of an East Coker villa.

Axminster. Vespasian also developed the road between the important Roman settlements of Dorchester (Durnovaria) and Ilchester to link with the Fosse Way, passing within three miles of East Coker.

There is evidence of significant Roman villas in both East and West Coker dating from the heyday of the Roman occupation. When excavated in 1753, the villa at Chesil Field in East Coker revealed mosaic pavements with Christian iconography and hunting scenes. The villa in West Coker revealed a bronze plate inscribed to the god Mars Rigisamus (which means 'greatest king' or 'king of kings') with a figurine of a standing naked male figure. The two villas are located unusually close to each other and form one of the few double Roman villa complexes in Britain. While the two agricultural units represented by these villas form the origins of the manors of East and West Coker and flax and hemp production in the area, it seems more likely that Saxon farmers would have been the first to grow these crops.

The land of villages and the villeins

The landscape historian W. G. Hoskins describes the Saxon period as being when 'England emerged as the land of villages'. These were blocks of countryside allocated to communities for maintenance by the authority of the king, the queen or a chief, with each area having a boundary defined by physical features (hills and streams), its shape governed by the agricultural and woodland requirements of the community. In reality, these boundaries do not seem to have changed materially through time. They became the vills, or manors, of Domesday. The vills were, as Hackwell (1957) notes, 'occupied by the villeins – the shepherd, the herdsman, the weavers, the thatcher, the baker – whose huts and cottages became the villages and subsequently the parishes as we know them today'.

It was not unusual, however, for the villages to be sub-divided into two or three parts when a number of different centres, or nuclei, developed. This was the case in Cocra, or Cochre, which was originally held under a single lord, eventually to become the North, West and East Coker we recognise today on road signs. A storied past. The undivided became divided.

Despite originally being part of the same manor, the three villages have always been, and still are, physically discrete entities separated by open fields and woods. Their 600-year-old administrative boundaries remain intact. The two parish councils (East Coker with North Coker and West Coker) function dutifully but separately. At the peak of its output of rope and canvas manufacture in the mid-19th century almost 5,000 people lived in the two parishes. Today, there are less than 4,000 residents. The creeping suburbia from an expanding Yeovil (once the centre of glovemaking in Europe) has, so far, been held at bay. Recent land grabs have been hotly contested and rebutted, especially a brutal development proposal in 2010.

Slaves and freemen

From 1016, the Saxon-owned lands in Somerset were held by Countess Gytha Thorkelsdóttir (c. 997–c. 1069), a Danish noblewoman, wife of Godwin, Earl of Wessex, and mother of King Harold Godwinson and Edith of Wessex. According to South Somerset District Council, 'The royal status of these lands [in Saxon times] derived from their worth as highly flexible and fertile agricultural properties and position at the cross-roads of the main North South and East West land routes in Southwest England'.

Although a new bishopric for Somerset built a wooden church in East Coker on the site of the current church of St. Michael and All Angels' in 909 AD, the name Coker first appears in the *Domesday Book* following the Norman invasion of 1066. In it, what is now East and West Coker appears simply as 'Cochre', under the heading: 'here are entered the Holder of Lands in Summersete'. It also appears in the Exeter Domesday statement with regards to 'the King's lands which Earl Godwine, and his sons hold in Sumerseta'.

Above: The 17th-century Grade II* listed Pavyotts Mill.

So, the King held Cocre with its population of 42 smallholders, 35 villagers, 7 slaves and 4 freeman, giving a total combined population of c. 200 (excluding women and children), owning 48 goats, 150 sheep, 20 pigs and 3 cattle plus woodland, meadows, 15 ploughs and a watermill.

The next official mention of Coker occurs early in the reign of William Rufus (1087–1100). The *Anglo-Saxon Chronicle* of 1429 notes that in 1088 some of the lands in Coker were granted to the Abbot of Giselbert of Limburg (Belgium) by the King. There is no further mention of Coker in official records through the reigns of Rufus, Henry I or King Stephen, from 1135–1154. However, early in his reign (1154–1189), Henry II granted lands at Cokre/Cochra/Kokre to the rebellious Baron Geoffrey de Mandeville,

Earl of Essex. In 1202 de Mandeville de Coker (as he is now known) paid two marks in scutage (a fee in lieu of military service to the then monarch, King John) for land at East Coker and West Coker – the first time that the two villages were named separately.

The de Mandevilles de Coker were to retain an interest in both the villages throughout the reigns of Henry III and Edward I until 1307, after which Hugh de Courtney, Earl of Devon, is recorded as inheriting East, West and a South Coker.

From 1590, the Portman family took ownership of the Manor of West Coker for the next 350 years. They encouraged a more diversified landownership in West Coker, allowing more yeoman farmers to own land and develop grand houses than was to be the case in East Coker, where the Helyars were Lords of the Manor. Ironically, the Portman family had been tenants of the Helyar Estate at Pavyotts House, Farm and Mill. Located on the Coker Water to the north-east of East Coker, it is acknowledged as an important property in the history of flax production. William Portman, who inherited the Manor of West Coker in 1646, became a prominent member of the Royal Society, a body that was to champion and sponsor the scientific work of William Dampier.

The arrival of the Helyar dynasty

In 1617, the land of the Manor of East Coker passed into the ownership of Archdeacon William Helyar (1559–1645), who was Elizabeth I's chaplain and held many ecclesiastical appointments, including Canon of Exeter Cathedral and Archdeacon of Barnstaple.The estate, purchased from Sir Edward Phelips, remained in the family's ownership for 334 years until 1850, with Coker Court, today a Grade I listed building, being the base for the Helyar dynasty. Coker Court (built adjacent to the parish church) was the Helyar manor house, containing the finest 15th-century great hall in Somerset and a fine Tudor wing. In the early 20th century, with the house in the ownership of Helyar descendent Dorothy Walker Heneage, Queen Mary was a frequent visitor, and it is believed that Queen Elizabeth (the Queen Mother) spent the first night of her honeymoon at Coker Court.

Archdeacon Helyar was a contemporary of Sir Walter Raleigh, the Elizabethan adventurer and explorer who retired to Sherborne Castle, located just 15 miles north-east of East Coker, in 1592. Being residents in East Coker also allowed the Helyar family to be close neighbours of Sir Ralph Horsey, who had supported Raleigh's projects in the West Indies relating to sugar plantations and the slave trade. Archdeacon Helyar would, therefore, have had first-hand knowledge of the Elizabethan zeal for exploration and the early colonisation projects.

Indeed, it was Archdeacon Helyar who initiated the family's acquisition of estates in Jamaica and, ironically, it was to be £600 raised from his lands on the island that was invested to complete the building of the Helyar almshouses at East Coker that he had initiated in 1640. The building work was interrupted by the plague of 1645, and the English Civil War and the almshouses were not completed until 1660, by which time the Archdeacon was dead and the work was completed by his grandson. In 1868 the almshouses, housing 11 women and 1 man, had an income of £46 per annum (equivalent to £4,403 in 2023), and the Grade II listed building is still operated as a charity.

Jamaican plantations, enslaved labour and the Coker Circle

Archdeacon Helyar lost his son Henry in 1634 and was succeeded in 1645 by his grandson, also called William Helyar, who inherited the Manor of East Coker. William Helyar the younger was a colonel in the Cavalier army, in charge of Royalist troops in Exeter at the time of its capture by Parliamentarian forces in 1646 during the English Civil War (1642–1651). He was tried for 'Treason to the Parliament' but discharged on payment of a sum of £1,522, with a Pardon in 1648 allowing him to drive the family's interest in developing their investments in the Caribbean forward.

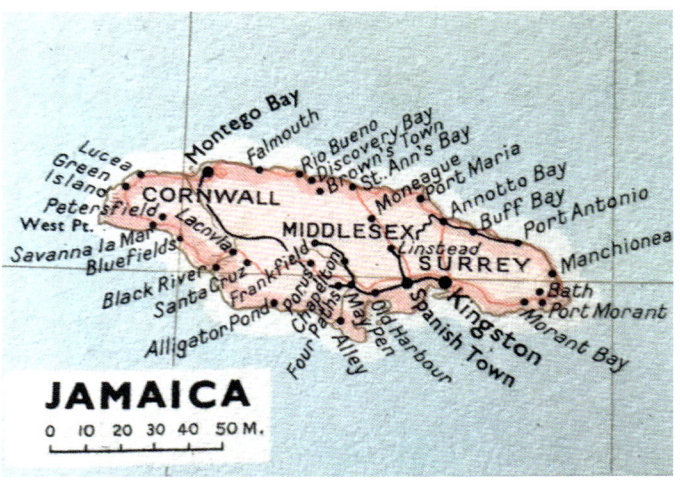

Above: Map of Jamaica, c. 1948.

In the mid-18th century, Jamaica became Great Britain's most profitable colony in the Americas. Not coincidentally, it was to become the colony that exploited slave labour most aggressively. One estimate suggests that at that time 90% of the population of 150,000 people lived in bondage, making their enslavers exceptionally rich. According to Vincent Brown, writing in *The Guardian* (March 2024), 'the island society became tumultuous with frequent uprisings, some of which threatened the very foundations of Imperial power, notably Tacky's Revolt in 1760'.

It was to be Colonel William Helyar who, together with his brother Cary, invested in establishing the Bybrook Estate as an integrated plantation in Jamaica. In August 1664, Cary Helyar wrote a letter to his elder brother, William. In it, he declared his intention to embark on what was to be his final adventure: to settle in Jamaica and carve out a place for himself and his family in England's burgeoning Caribbean economies. 'Here [the Spanish-controlled island of Tenerife] is a ship bound for Jamaica,' he wrote, 'upon which I intend God willing to imbarque & there settle … for I am weary of rouling, I will therefore fix & grow mossy although it bee upon my skull.'

William agreed to organise the shipping of supplies from England whilst Cary, who arrived in Jamaica in autumn 1664 at the age of 30, was to acquire and develop the land on the island. He was not a patient man. Whilst he followed the general patterns of developing a plantation, he accelerated the process, making poor business decisions along the way.

This was a period still referred to as a 'raucous decade of English Jamaica', still wrapped up in wars of plunder and piratical attacks by the likes of the infamous Welshman Henry Morgan – indeed, Anglo-Jamaica was known as a haven for piracy. Jamaica was located in an important strategic location for the English. It was England's closest possession to the Spanish mainland and thus opened up new possibilities for legal and illegal trading opportunities with the Spanish.

Cary was convinced that he had a businessman's eye to the future. Operating as Cary Helyar and Company, he had spent the previous three years trading enslaved people and exporting 'wines, sugar, logwood and elephant's teeth' and other Jamaican produce with Spanish ports in the Caribbean and beyond. With this background, Cary believed the family could make their fortune.

Over the course of the next four decades, the Helyar family remained the owners of what became Bybrook Plantation and navigated the ups and downs of the sugar estate. During their tenure, an eventful 44 years, there were births, deaths, marriages, mistresses, lawsuits, profits, debts, betrayals and even murder by a dung fork as the family tried to carve their place in Jamaica's emerging sugar industry. It all began when, in 1669, Cary Helyar began acquiring land for the sibling's joint venture in 1669, the original intention being to focus on growing cacao. However, following a blight in the crop in 1670–71, they turned their attention to growing sugar cane.

Brothers Cary and William Helyar initially purchased 160 acres of uncultivated Bybrook land and then were granted permission to buy a further 466 acres of land adjacent to the well-established plantation of Sir Thomas Modyford (the Governor of Jamaica and former officer with William Helyar in the English Civil War) at Sixteen Mile Walk in St. Katherine's parish in 1669 (now known as St. Thomas-in-the-east). Modyford became Cary's mentor and was at the heart of what became known as the Jamaican Circle. Cary was convinced by Governor Modyford to purchase the land for Bybrook. It was an expensive undertaking for both brothers, but as partners they agreed to individually assume one half of the costs associated with the construction of the plantation and split the profits evenly between them. Moreover, the Helyar brothers viewed Bybrook as a long-term investment, but as we will see, the adventure did not end well.

The Coker Circle

Though Cary and William knew little of planting when they acquired the enlarged plot in St. Katherine's Parish, Cary in particular was certain that they could not build a plantation alone. Therefore, in its first years, Cary used his social networks to amass the necessary knowledge, financial capital and labour in order to create a viable plantation. One was centred on the family home in East Coker, and was known as the Coker Circle, while the other was centred locally, the Jamaican Circle. Both of these networks were uniquely influential on the settlement of Bybrook. While the Coker Circle functioned as a means of financial and logistical support for the fledgling plantation, the Jamaican Circle formed the social and political basis for Cary's life in Jamaica.

According to Daisy Ramsden's 2015 research into the moral economies of Anglo-Jamaica, 'William's influence within the Coker circle was keenly felt by all involved, no doubt due to his role as Cary's equal partner in Bybrook. As his equal partner in both debt and profit, it was in William's best interest to supply his brother's needs in a timely manner. In fact, it is hard to overvalue his impact on the development of the plantation, as William's impact on its earliest years was second to none. As the only permanent resident of England in this story of the divergence of Jamaican values from English ones, he functioned as a sort of litmus test for English values – he was immersed in and was typical of the English economy of obligation. His judgement of an individuals' character carried significant weight in the business affairs of Bybrook. William Helyar controlled both Bybrook's purse strings and its English labour supply, which meant that if he deemed an individual to be of bad character, he simply refused the request for more financial credit or labourers.'

Above: Coker Court, East Coker.

The Register of Baptisms for St. Michael and All Angels' Church also indicates some potentially unintended consequences of the Coker Circle's links to Jamaica. In 1662 the register records the baptism of a baby girl called Hagar – the name of an Egyptian slave who bore a child for Abraham in the Book of Genesis. Thirty years later, there is an entry dated 25 May 1692 of the baptism of Thomas Wm. Helyar Esq. Black (a Moore).

No place for ugly weavers

Whilst Cary Helyar followed the general principles for developing a plantation based on established knowledge, he accelerated the process. As a result, he struggled to find the skilled labour needed in Jamaica and put pressure on his brother and the Coker Circle to solve this problem. On several occasions he wrote to his brother asking for artisans to be sent from Somerset, including carpenters, masons, bricklayers and potters, but, according to his letter of 4 June 1672, 'no ugly weavers'.

For William, the solution was to persuade his local indentured labourers to work at Bybrook, offering them up to 30 acres of land in Jamaica upon the completion of their indenture. William looked no further than the immediate area around the Cokers for this source of labour and five families accepted the offer.

The term 'indentured servant' refers to a system for financing immigration to the Caribbean, primarily during the colonial period. Many in the West Country who could not afford passage sold themselves to merchants and seamen in exchange for transportation to the colonies. This arrangement was spelled out in a contract, called an indenture, in which the emigrant agreed to work without compensation for a fixed term, typically four or five years. Generally, the servants often entered into such contracts freely, but in many cases were persuaded by their landlords, as appears to have been the case with the Helyars, to commit to an indenture. Labour shortages in the colonies enabled indentured servitude to flourish until increased African slave imports during the 18th century triggered its decline.

Here are the names of three locals who we know became indentured servants. On 6 February 1685, John Pooler of Yeovil, Somerset, was indentured for seven years to John Napper to serve in Jamaica and sailed on the *America Merchant*. On 2 May 1685, William Fivian of Yeovil, Somerset, was indentured for seven years to John Napper to serve in Jamaica and sailed on the *Dragon*. On 4 May 1685 Walter Summers of Preston (Plucknett) was indentured for four years to Thomas Kirke to serve in Jamaica and also sailed on the *Dragon*.

In addition, one of these indentured Somerset labourers named by Cary was Henry 'Goodman' Hodges, who was encouraged to enhance his own reputation by influencing others from the Cokers to be part of the Bybrook adventure. Indeed, in May 1671, soon after the Hodges family arrived, Cary wrote to William, stating, 'Goodman Hodges tells mee if you meet with The Old Croker of Coker and tell him of his [Hodges] being here that hee will help you to enough tradesmen.'

In their first year, the Helyar brothers bought 12 enslaved people and hired a small group of white servants. Upon the sudden death of Cary in 1672 the Bybrook Estate had expanded to 1,236 acres with their joint investment of £1,858 on the property, of which £1,205 was spent purchasing 55 enslaved people – twice the price quoted by Cary to his brother. By this time only 24 acres (less than 2% of the land) had been planted with sugar cane and none harvested.

Cary's death and burial marked the beginning of William Whaley's tenure as manager of Bybrook. Whaley, who was from London but was William Helyar's godson, had written to his godfather volunteering to 'goe to prentice off to Jamaica' and had journeyed to be part of the Bybrook Estate in 1670. Cary bequeathed his share to their new manager, who immediately conveyed it to William Helyar along with Cary's debts. William, the absentee owner, soon realised that Whaley was heavily dependent upon the connections that Cary had established as part of the Jamaican Circle, but that this reputation was being rapidly eroded.

Faire promises and a seasonable offer

William Helyar made the bold decision to send the orphaned son of one of his East Coker tenants, a young William Dampier, with 'faire promises' to take on a meaningful role in the management of Bybrook and to be his eyes and ears on the activities of William Whaley in Jamaica.

As a boy, Dampier had impressed the squire with his knowledge of the crops grown by Helyar's tenants. Dampier wrote proudly, '[I] came acquainted with them all, and knew what each sort would produce ... in all which I had a more than usual knowledge for one so young, taking a particular delight in observing.'

So, on 6 April 1674, the merchant ship *Content* sailed down the Thames, with the undoubtedly nervous, thin-faced, 22-year-old Dampier on his way to work on the sugar plantation. According to Diane and Michael Preston, Dampier had baulked at the last moment for fear of being sold as an indentured servant, as was the case with others from Somerset, when the ship berthed in Jamaica.

Above: Chur Lane – one of the sunken lanes or Coker tunnels – West Coker.

The Prestons wrote in 2004 that 'As wind filled the sails and the small vessel began to creak and roll, it was too late to change his mind. However, he had sensibly agreed with the ship's captain, John Kent, that he would work his passage as an able seaman. As such, the law required Kent to discharge him as a free man on his arrival.'

William Dampier's position was, he soon realised, ambiguous, the exact nature of his arrangement with William Helyar – 'a seasonable offer' – being unclear. Helyar had provided Dampier with some supplies on the understanding that he would work for a period in return, but there was no formal agreement between them specifying for how long or on what terms. This perplexed Helyar's London agents, Rex Rock and Thomas Hillyard, particularly when Dampier began to demand further items before the voyage: paper, ink and quills, a pair of shoes and a pound of soap. He also insisted on a grater, a nutmeg and two pounds of sugar-essential ingredients for making punch. The two men complained to Helyar, saying, 'that William Dampier has been very extravagant.'

Not long after his arrival at Bybrook, Dampier was reporting to his sponsor that Whaley was refusing to fulfil his obligations to him and also highlighted the perceived mistreatment by Whaley of several white labourers (fellow former residents of East and West Coker). To all intents and purposes this letter from Dampier to William was his resignation. Whaley contested these accusations in turn, doing his best to undermine Dampier's reputation and relationship with Helyar and saying that 'had he bin any thing ingenious hee might almost bin a good boyler, but he thought it undervalued him.'

Dampier was not the only one to desert the Bybrook Estate, but Whaley, far from being concerned by the loss of white workers, saw it as a cure of ills. Writing to Helyar in 1674, he described the exodus as 'by a company of wast full people' and calling the mistress of the plantation's doctor 'a whore' who was 'the nastiest wasting shit as ever came in to a house.'

Beyond the inadequacies of his manager in Jamaica and the departure of his spy, William Helyar had more issues to contend with, not least the legal claims from both his mistress (and mother of his son) and his brother's widow, Priscilla, not only for money owed to her but also for her claim on an enslaved labourer that she considered to be her inherited property.

On the death of Whaley in 1676, the output of the Bybrook Estate was valued at £2,737, excluding the value of the land and the sugar works, with 104 enslaved people. With a total value of £4,000 (equivalent of about £800,000 in 2024), 'it had 80 acres in sugar acne and an exceptionally large sugar works with a water mill, six coppers in the boiling house, three stills in the distillery, and five hundred sugar pots in the curing house. No other Jamaican estate inventoried between 1674–1678 possessed so many slaves and servants.' Whaley did not manage any livestock, and as a result was always asking his brother to send more supplies from Somerset. Whaley was also not managing the sugar crops very efficiently, as one report noted that 'a plantation of this strength should have turned out more than the forty-six hogheads Whaley says he made in 1675.' Indeed, just 13 years later, with the estate managed by William's son, John, the production had increased to 241 hogheads of sugar – the outcome of the work of 144 enslaved people – however, the land had been poorly managed and the soil was 'quite wore in out and worth very little', meaning the whole enterprise sold for £2,350.

Bybrook, on any reckoning, was a failure for the Helyar family, despite their constant investments. They sold it without amassing any significant profit and certainly not the riches envisaged by Cary. There was to be no family legacy for the business and there was the loss of reputation within Anglo-Jamaican white society and with the Jamaican and Coker Circles. Poor decision-making by ineffectual managers, together with Cary's affair, his illegitimate son, the way he deceived his brother William in underestimating the costs of the business and exaggerating the profitability, all conspired towards the

Above: Heritage tree, East Coker.

failure. The problems persisted. William's son John was embroiled in a well-known affair with Betty, an enslaved labourer who was the mother of Thomas, his son, both of whom he allegedly refused to free when he left Jamaica.

Throughout this period of the Helyars' interest in trade in the Caribbean they would have been highly dependent upon sailing ships and understanding of the importance of the sailcloth and ropes made on their Coker estate by the 'ugly weavers', relying upon and exploiting their indentured labourers in the Cokers to help drive their Jamaican enterprises. Back in Somerset, as we will see, they actively encouraged their tenants to grow flax and hemp, they allowed looms in their tied cottages, they developed mills and yarn bartons and they nurtured enterprise and innovation, their motivation undoubtedly being more mercenary and pecuniary than benefaction. The sailcloth and twineworks serviced their exports and import activity whilst their tenants, through these activities, were able to pay the landlord higher tithes and rents.

Above: Sunken lane, North Coker.

Open fields cut by sunken lanes

Today, each Coker village seems comfortable within its landscape, located in a vale on the Yeovil Sands with its well-tended verdant parklands, pastures, winding hedges, orchards, tree-lined streams wandering through pastures, meadow, moor and marsh. The open fields are sprinkled with lonesome specimen trees, such as the giant smooth-leaved elm (*Ulmus minor*) that was recently proclaimed as the finest free-standing specimen in Europe. It stands sentinel over the East Coker Cemetery. This intimate landscape is encircled by gentle hills, the ground rising out in a radiating pattern that forms the Yeovil Scarplands and consisting of Jurassic-era sandstone, with harder and more calcareous Forest Marble rocks forming the ridges of Coker Hill, Windmill Hill and Abbot's Hill, capped with broad-leaved woodlands.

They say poets make the best topographers, and T. S. Eliot lived up to this claim in his eponymous tribute to his ancestral home. He describes that when you arrive in East or West Coker you thread your way through the deep, hidden, sunken lanes – known as the Coker Tunnels – in lines immortalised in the second of his *Four Quartets, East Coker*:

> 'In my beginning is my end. Now the light falls
> Across the open field, leaving the deep lane
> Shuttered with branches, dark in the afternoon,
> Where you lean against a bank, while a van passes,
> And the deep lane insists on the direction
> Into the village, in the electric heat
> Hypnotised. In a warm haze the sultry light
> Is absorbed, not refracted, by grey stone.
> The dahlias sleep in empty silence.
> Wait for the early owl.'

This ancient, complex network of narrow, hollow lanes is entrenched in the Yeovil Sands – a silty, fine-grade calca sandstone, the result of natural processes as the fine loose sand has been easily washed away over the centuries. The ground was further impacted by human activity as the passage of smugglers' carts and farmers' wagons, together

with the pounding of livestock, continually eroded the bedrock. The lanes form a maze of byways whose routes defy 21st-century logic but whose names give us clues as to their story – Culliver's Grave, Dibbles, Halves, Stoney Lane, Back, Long Furlong, Gunville, Foxholes, Moor and Chur Lanes. At the bottom of Chur Lane, the Coker Rope and Sail Charities are now planning to restore the old West Coker Village pound.

Top: St Michael and All Angels', East Coker.
Left: The Helyar almshouses, East Coker.

The Place

Flax and Hemp – The Essential Raw Materials

On Thursday evenings from July 1958 and for the next seven years, the Bush radio in Lea-on, our compact thatched cottage in East Coker, was tuned to Kenneth Horne's radio show *Beyond Our Ken*. The programme featured a fictitious gardener called Arthur Fallowfield, played by Kenneth Williams.

His response to any question was always, 'The answer lies in the soil'. The character was a parody of our highly respected local gardening expert, Ralph Whiteman, whose favourite response to a listener's enquiry on the BBC's long-running *Gardeners' Question Time* in the 1950s was generally 'the answer lies in the soil'. Whiteman, a Dorset farmer's son from Piddletrenthide, recognised that the soils around the Cokers were derived from a similar geology to those of West Dorset, soils he knew allowed flax and hemp to thrive – what Eliot would describe as 'the significant soil'.

The importance of the geology and the soil to the development of the sailcloth industry in the Cokers is a recurring theme that was also recognised by William Dampier. In his *Journal* of 1689 – later to form part of his *A New Voyage Around the World,* published in 1697 – Dampier makes an important observation about the soils in Sumatra and reflects on those around the Cokers. He writes: 'I took as much notice of the difference in soil as I met with ... having been bred in my youth in Somersetshire, at a place called East Coker near Evil [Yeovil]; in which parish there is a great variety of soil, as I have ordinarily met with anywhere, viz. black, red, yellow, stony, clay, morass or swampy. ... I had more reason to take notice about these things, especially because this village in great measure is let out in small leases for lives under Colonel Helyar, Lord of the Manor.; and most if not all of these tenants had their own land scattering in small pieces so that everyone had some piece of every sort of land.'

Dampier continues, 'My mother being possessed of one of these leases and having of all these sorts of lands. I became acquainted with them all, and knew what each sort would produce viz, wheat, barley vetches, flax, or hemp.'

In 1795, a review of agriculture in Somerset also reported that the 'rich fertile country through Yeovil to Crewkerne including the villages of Hardington, Pendomer, Closworth, Odcombe, but mainly in the Cokers, allows flax and hemp are cultivated in abundance.'

From fibres to textiles

Flax is a plant first brought into cultivation in the Near East (Fertile Crescent) by 8,400 BC. It spread northwestward, reaching Central Europe by c. 5,000 BC and Britain by

around 4,000 BC. Two main uses for flax have been recorded in archaeological, historical and ethnographic sources: first came the use of linseed and linseed oil for cooking, of which there is evidence at many sites across Britain dating back to the Neolithic period. The earliest known evidence of the use of flax for textile production dates from the (now) Republic of Georgia 30,000 years ago, and the Romans would later use the fibres for sailcloth.

Since at least the 13th century, or maybe as early as the Roman period, the area around the Cokers has been known for the growing of flax (*Linum usitatissimum*) and hemp (*Cannabis sativa*). Flax, with its June pale blue flowers, thrived on the deep, organic-rich marl and clay soils of the Middle Lias, whilst hemp flourished on the free-draining, fine-grained, soils on gentle slopes of the outcrop of Bridport Sands (locally known as the Coker Sands) running between Beaminster and Yeovil. Even today, flax can be seen as blue swathes across the south Somerset countryside in late spring. The delicate blue flowers are supported on thin, reed-like, stems which are harvested for making twine and textiles whilst the round seed head bolls are used for flour and oil.

In the Cokers, the flax, which was sown in March or April, could be harvested for its fibres by pulling out the whole plant in August while some plants provided seed for the coming season. However, in the 19th century the Cokers were supplementing the local sources of seed by using seed imported from Riga, today the capital of Latvia but at the time an important Baltic Sea port of the Russian Empire.

Hemp seeding took place in May. Being dioecious, the male crop was harvested after flowering in August whilst the female crop was left until seeding finished in September for the use the following season. In 1808 a letter by Henry Saunders of Bridport noted that 15 women will draw one acre of male hemp per day at a cost, including liquor, of £1 2s 6d and because there was 'no machine for dressing hemp; the process is performed by women and children who scale it by hand and are paid a 1d per lb for the male hemp for which the manufacturer buys it at about 16s 6d per weight'.

Located just 20 miles to the south, Bridport, with its harbour at West Bay, features significantly in the story of the Cokers. Since the Middle Ages, Bridport has been associated with the production of sailcloth, rope and nets. The earliest official record of this industry dates

---

Left: Blue and white flax (*Linum ascyrifolium*) with pale blue flowers, a native of Europe.

Steeping in Penyes Pitts

There is further indication that the Cokers' interest in the manufacture of yarn, twine and sailcloth was long established before its heyday of the 1700s. In 1309 hemp formed part of the tithe to the Rector of West Coker. Parish records for 1358 tell us that the Palmer family was producing yarn from Somerset hemp and in 1438 that hemp and flax was grown in cottage gardens as well as in fields and yards. In 1503, Glastonbury Abbey is noted as owning Penyes Pittes in West Coker, ponds for the steeping (or retting) of fibres – a process of placing cloth and fibres in water to remove impurities, preparatory to bleaching. At this stage it was assumed that, because most sailcloth used in England at this time was imported from France, the products were yarn and twines for local use.

It is no surprise, therefore, that by 1692 it was noted that 'great quantities of flax are cultivated in the village for making cloth, ticking, sewing thread, cabulettes and hausers'. However, by 1798, during the reign of George III, a bounty of 3d was provided on every stone (14lbs/6.35kgms) of hemp and 4d on every stone of flax produced in order to boost production in England. Grants were made available, with 15 producers living in West Coker and 5 in East Coker applying for this support and over £186 being paid out to villagers John and James Wadman and Elias Hawkins, with Joseph Rendall being the largest grower.

Once the flax and hemp fibres had been purchased from the growers by the merchants and carefully weighed, they were then allocated for manufacture and distributed to different artisan workers, who tended to operate in family units. The men undertook the initial dressing of the crop and the women spun the yarn, then made twine, line and rope. The completed products were then returned to the merchant, who paid for the added value and would check the weight returned to ensure it was within an agreed limit.

Pulling, rippling, scaling and hackling

Processing the raw materials from flax and hemp involved a series of specialist activities, some of which were mechanised by the mid-19th century.

from 1211, when King John ordered that Bridport make 'as many ropes for ships both large and small and as many cables as you can'. He also requested a thousand ells of cloth for making ships' sails. The raw materials needed, flax and hemp, were readily available in the surrounding countryside, including the Cokers. Accounts for from 1309 showed 30 sheaves (sheafs) of hemp being transported to Bridport from West Coker whilst accounts for 30 May 1356 name 14 producers (7 from East Coker and 7 from West Coker), who supplied 1,220lb of *filum cannabi* (hemp) to Bridport. This suggests a long-established relationship between the port and the villages.

Above: Growing flax at the Dawe's Twineworks as part of the OSR Arts Project.

Above: Flax pulling in fields near East Coker; water retting at Slape Mill; scutching and scaling at Bunford Flax Mill; hackling at Pymore Mill and sorting at Slape Mill.

Flax and Hemp – The Essential Raw Materials

After pulling the plants from the ground, the initial scaling took place, which involved women and children stripping the stems of all foliage. 'Rippling' then took place, using a coarse comb which would remove the seeds. Before the plants could be used for textile production, the soft, lustrous and flexible flax fibres, from the fibrous bast beneath the surface of the stem, were extracted through a process called 'retting'. This consisted of soaking the flax stems until they partially broke down and the fibres could be removed. The stems were soaked for 15–20 days using naturally occurring dew in open fields or in special pits, retting ponds, ditches or slow-moving streams. Following drying, the fibres could be removed from the stalks.

Dressing was the all-encompassing term covering these processes of bolling or breaking (to help the removal of any unwanted woody elements), scutching (the removal of the unwanted elements), then heckling, or hackling, to comb out and separate the fibres aligning them in the same direction ready for spinning. Spinning converted the weak fibres into a stronger yarn and subsequent twisting of two or more yarns produced the twine. Spinning originally took place in the home using a simple spinning wheel, but later it was done on a large scale in spinning mills.

Alchemy – the key to success

The most important aspect of preparing the fibres and the key to the remarkable success of Coker Canvas and its superiority over all others was the bucking (or boiling) of the yarn in a bucking house. This was usually a single-storey building close to running water containing furnaces and cisterns for boiling the yarn. The yarn was initially steeped in hot water, then bucked in a special local recipe of an alkaline lye and potash. The yarn then dried on grass in bleaching fields, or on wooden rails in yarn bartons, for two to three weeks, then these processes were repeated four to five times before the yarn was finally soured with milk for about three weeks to neutralise it. The later introduction of chlorine as a bleaching agent in 1854 sped up the process.

As we have seen, at the heart of the industry was the manufacturer, who would organise for the crops to be grown and harvested and for the raw materials and finished products warehoused (he would have one or more bucking houses). Once the yarn had been produced, the manufacturer would then organise the distribution of the yarn to the spinners and weavers or, with the arrival of power looms in the 1850s, it would be taken to be woven in their own factories. The manufacturers would then deal with the sale of products directly with the merchants.

Weaving involved the interlacing of the warp (the longitudinal threads) with the weft (the lateral threads). This job originally took place in the home using a hand-operated loom which could weave a 24-inch-wide cloth. Twine or rope manufacture was a separate technique that developed when twine making took over from making sailcloth. It involved spinning two or more strands of yarn using outdoor twine or ropewalks, the open strips of land (later roofed to give protection) that were sometimes over 300 yards long.

A dramatic increase in demand

The demand for yarn dramatically increased as larger sailing ships were built and both the merchant and naval fleets increased in numbers. Despite a bounty being introduced to boost local production, supply could not keep up, leading to merchants and manufacturers turning to developing import trade with Russia and northern European countries. In 1796, 9,108 tons of flax were imported from Russia.

For the next 200 years, local farmers continued to produce flax and hemp to satisfy local demand for sailcloth and twine. Their fortunes ebbed and flowed with the vicissitudes of the marketplace for these products.

The next major external intervention to stimulate the growing of these plants came at the turn of the 20th century. This was the Development Fund Act (1909), which established experimental flax-growing centres across the UK, starting in 1911, in order to boost homegrown

Above: Land girls helping with the flax harvest at Longlands Farm. The farmer, Jesse Crumpler, was managing director of Wessex Flax Factories Ltd.

production and reduce the dependency on Russian imports. Incentives were provided by the Government to support flax growing by farmers – not only was the seed provided free of charge, but growers were paid a guaranteed £14 per acre for their whole crop (seed and straw) as well as £8 10s for higher quality products. It was estimated that a well-managed crop could produce two to three tons per acre. The *Kelly's Directory* for this year for East Coker stated that the population was 731 and 'The chief crops are wheat, barley, beans, roots, flax and hemp with large amounts of pasture.' For West Coker, the population was 871 and 'most inhabitants are chiefly employed in the manufacture of twine and glove making.'

In 1915, the University of Leeds took the lead in the research and development of flax experimentation stations. The *Journal of the Textile Institute* records that the university was also promoting the formation of the British Flax and Hemp Growers' Society, 'which controlled a flax experimentation station in Yeovil in Somersetshire'.

From flax field to factory

This experimentation factory was known as the Bunford Flax Factory. Located on the West Coker parish boundary with Yeovil, it was constructed by the Government's Flax Production Branch of the Ministry for Agriculture and Fisheries (founded in 1917). In addition, a flax-growing residential camp for the volunteers, established at Barwick House, near East Coker, courtesy of the Messiter Estate, was thriving. Hundreds of women volunteered through the Women's Land Army and came to the Yeovil area to work in the flax fields. At its peak, the camp at Barwick House was the temporary home to over 600 female flax pullers.

A postcard sent in July by a flax puller based at Barwick House to her family in London stated that 'flax pulling is rather dull, but the country is so lovely that it makes up for it.' She noted that her pay was 1s per day and there were 75 bell tents as their accommodation. Another postcard, sent by a flax puller at Barwick House in August, said that 'next week the camp will close. There are only about 100 of us left now'.

Above: Yarn winder at Dawe's Twineworks.

At this stage, the flax was mostly turned into linen and used to cover the wings of biplanes. It was no coincidence that the Westland Aircraft Works, founded in 1915 as a division of the local engineering firm, Petters, was built at West Hendford, close to the Bunford Flax Factory and just three miles from Barwick House. The new company was contracted to build aircraft that needed extensive amounts of linen to cover their wings. Orders were originally received to construct under licence Short Type 184 seaplanes. Orders for other aircraft soon followed during the First World War.

Demand for flax was heavy, with 180 yards of linen needed for each plane, and it needed replacing every 12 flights, although some reports claim that, depending on conditions, some of the linen used could last for 5-10 years. The *Annual Report for the Flax Production Branch of the Department for Agriculture and Fisheries* recorded that the total area of land in Great Britain contracted to grow flax for use in its own factories was 12,382 acres and that 26,290 bushels of seed was made available for this purpose. The cost of harvesting was estimated at £10 per acre plus £1 per acre costs for the camps, food and water. A set guaranteed price was approved at £14 per acre. The amount produced by each of the Ministry's centres for 1918 was:

Centre	Total tons	Cwt. per acre
Fife	2,750	41
Suffolk	3,260	37
Selby	6,300	48
Yeovil	6,700	36
Peterborough	7,500	48

In 1920, the Ministry of Agriculture and Fisheries decided to dispose of the Yeovil 'centre' of flax factories (at Bunford, Preston Plucknett, South Petherton, Lopen, Taunton, Bridport and Dorchester). All were sold to the investment bankers A. Mitchelson & Co. Ltd. The Bunford Flax Factory was purchased by Wessex Flax Factories Ltd.,

whose Managing Director was Jessie Crumpler, a farmer who owned Longlands Lane Farm, North Coker.

Four years later, the Bunford Flax Factory fell into liquidation, leaving 400 flax growers in the area unpaid for contracts agreed in 1922. Two years later, the Bunford receivership leased the factory to the Linen Industry Research Association, and it was operated by the Flax Industry Development Society Ltd., who appointed none other than the hapless Jessie Crumpler as its Managing Director. However, they were unable to provide the bulk supply of the flax seed. As a result, the Flax Industry Development Society Ltd. was formed in 1926 as a not-for-profit organisation, funded by the Government. Throughout the late 1920s the factory widely advertised to 'receive offers of land for the flax crop'. However, within 10 years the Flax Industry Development Society Ltd. was dissolved, with the Bunford Flax Factory closed and sold to Aplin and Barrett to become their Watercombe Creamery. It is now the Old Creamery furniture retailer.

Above: Yarn baling.

The Making of a Global Brand

Twine factories and ropewalks were commonplace in the Coker villages in the late 19th century, however, prior to 1850, it was sailcloth, Coker Canvas, which was the mainstay of local business. Initially, this was a domestic industry, with the spinning of the yarn and then the weaving undertaken in cottages. These home-based activities were replaced in the 1850s by factory-power loom weaving.

Sails for the centuries

Sailmaking goes back thousands of years and was practised by many ancient civilisations, from the Pacific Islanders to the Egyptians. Originally made from hemp and flax (linen), then cotton and more recently synthetic polyester derived fabrics, sails have developed enormously over the years. Throughout, the single aim was to provide a robust and stable expanse of cloth that would create and keep a shape to harness the force of the wind to propel a boat.

For years it was thought that the introduction of the sailcloth industry in England dated to the late 18th century, however, research has shown that there was a native industry based in the west Dorset and south Somerset area at the time of King John in 1211, when there was a reference to an order for *'pro mille ulnis tele ad uela nauim facienda'* (for a thousand ells of cloth for making ships' sails – an ell being a former measure of length, equivalent to six handbreadths, used for textiles, locally variable but typically about 45 inches in England and 37 inches in Scotland. The word means 'arm' and survives in the form of the modern English word 'elbow').

Whatever the status in the early years of this indigenous and localised sailcloth industry, it went into decline in the late 14th century due to the import of sailcloth from Brittany. Reliance on French imports of sailcloth proved to be a precarious state of affairs, especially when needed to supply sails for Henry VIII's navy. About this time, England was seeking to acquire 6,000 pieces of sail canvas from French suppliers. On the basis that 20–25 pieces of canvas was needed per ship, the French Government correctly interpreted this order as reflecting an expansion of the English fleet and declined the order. As a result, the emerging conflict with France directly led to the revival of the English sailcloth industry.

About this time, we see the first records of the Rendell (sometimes spelt Rendall) family in the Cokers. Their lineage was to produce numerous characters over the

Above: The Coker Sail Cloth Work poster, c. 1900.

years that were integral to the Coker Canvas story. Their occupations over the centuries included: coachman, coal merchant, highway surveyors, pub landlord, stonemason, soldier, rock blaster, twinemaker and, of course, sailcloth makers.

In search for the perfect arte

During the 16th century, the domestic capacity for sailcloth production had to be increased and the industry reinvigorated. As a result, Breton sailcloth weavers were brought over to England to help develop domestic production and skills, with Ipswich developing as the main centre for sailcloth manufacture. It is not clear if any of the Breton weavers found their way to the Cokers. A 1553 Act of Parliament instructed that one quarter of an acre of hemp or flax had to be planted for every 60 acres of arable land across England. Local sailcloth production increased, and in 1565 there is a record of the *Jesus* sailing from Bridport (West Bay) with a cargo of local hemp derived from the surrounding villages.

The *Annals of West Coker* states that in the 1590s the 'perfect arte and skill of sailcloth weavinge' was introduced into England from France. As supply increased across the country, great variations in quality emerged such that in 1602 a further Act of Parliament was introduced to regulate the craft and to boost the production and quality of sailcloth. The 1602 Act of Parliament specifically addressed the 'unskillfullness and badness of sailclothe made in England', introducing the requirement that 'none shall make such cloth except such that as shall have been apprenticed [for seven years] and brought up in the trade … And such cloth shall only be made of hemp … And not of any other length than three and thirtie yards, nor any lesse breadth than three quarters of a yarde.'

As an industry that we would recognise, it came to prominence in the 15th century, particularly with the creation and ascendancy of the Royal Navy in Britain. By the late 16th century, West Country sailcloth was actually known as 'English Duck'. The Dutch word for cloth, *doek*, evolved into the English word 'duck' in reference to sail canvas, hence Duck Lane in West Coker. Duck was typically made from cotton or flax, with some use of hemp.

Above: Spinning wheel on trade token of Shepheard & Co., Plymouth, 1796.
Right: The Brett & Cayne token, 1797.

By 1634 the prestige of Coker Canvas seems to have been well established, as John Giles of East Coker (a tenant of the Helyar Estate and uncle and guardian of the young William Dampier) is recorded as making the sailcloth for King Charles II's *Sovereign of the Seas*. Another sailcloth maker, George Elrond, is also referenced at this time.

Above: *Sovereign of the Seas*, flagship of the English Royal Navy during the 17th century, with sailcloth made by John Giles of East Coker.

The wild sea must be her port

Sovereign of the Seas was to be the flagship of the English Royal Navy. She was ordered on the personal initiative of Charles I of England as a prestige project and a deliberate attempt to bolster the reputation of the English crown. The decision provoked much opposition from the brethren of Trinity House, who pointed out that 'There is no port in the Kingdome that can harbour this ship. The wild sea must be her port, her anchors and cables her safety;

if either fail, the ship must perish, the King lose his jewel, four or five hundred man must die, and some great and noble peer'. These objections were overcome with the help of Admiral Sir John Pennington, and work commenced in May 1635. During the reign of William III, *Sovereign of the Seas* became leaky and defective with age and was laid up at Chatham Dockyards for repairs late in 1695. She ended her days, in January 1696, by being burnt to the water line as a result of having been set on fire by accident. In her honour, naval tradition has kept the name of this ship afloat, and several subsequent ships have been named HMS *Royal Sovereign*.

Pepys and the Pleys

In 1665, the diarist Samuel Pepys reported, '[here] comes Cutler [a well-known merchant who supplied the Navy the victuals and goods from the Baltic] to tell us that the King of France hath forbidden canvas to be carried out of his kingdom.' This decision saw the necessity for the development of English sailcloth by the Government. As a naval administrator, Pepys was a crucial figure in the growth of the Royal Navy and was to become Chief Secretary to the Admiralty.

Almost immediately, Constance Pley, wife of George Pley Snr. (then the collector of customs at Weymouth), offered Coker Canvas to the influential Commander Middleton at 20s an ell. A lack of response from Middleton caused Constance to complain to Pepys that 'tis a pity to meet so great discouragement in so good work as this English sailcloth.' Constance continued with her advocacy, implying that cheaper, poorer quality cloth was being made and offered to the Navy from elsewhere as inferior imitations of the West Country, or Coker Canvas, pattern.

Constance and George Pley had a personal stake, as they directly employed many of the villagers in East and West Coker, who wove the strips of cloth on hand looms in their cottages. Coker Canvas was already regarded as being superior, whiter and demonstrably stronger than its competition. It was proving to be the canvas of choice for the Navy, with most of the production possibly heading for the dockyards of Portsmouth and Deptford.

By the mid-17th century, Coker Canvas was being used for ships sailing to the Straits of Gibraltar and the fishing grounds off Newfoundland. Then, in 1657, George Pley, who had been supplying the Navy with French canvas, sent three pieces of locally made canvas for the Navy's inspection. It was described as 'white and not so fair as the French' because it had been bucked in the yarn prior to weaving. This same process had been used in making Coker Canvas for almost 30 years and was preferred by local merchants, whereas other English and French canvas was bucked in the piece. The Navy's correspondence with the Pleys at this time acknowledged the fact that, in 1634, John Giles from East Coker had indeed supplied the sailcloth for *Sovereign of the Seas*.

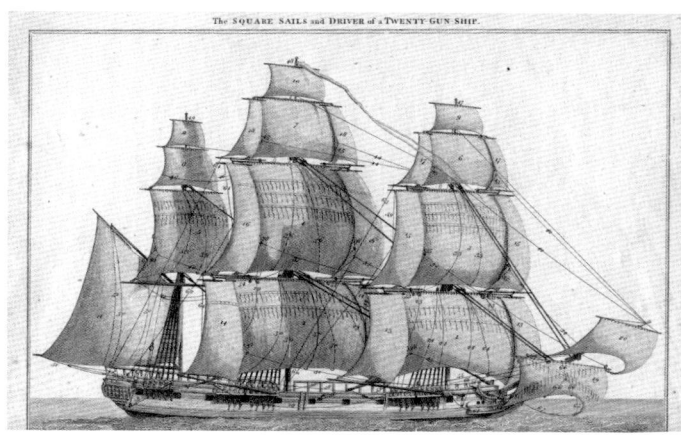

Above: The square sails and driver of a 20-gun frigate would have c. 250 bolts of canvas, while first-rate ships, such as HMS *Victory*, would have 450 bolts in a suit of sails.

Quality over quantity

The Navy was slow to respond to George Pley's letter, although it did admit that the sample had been passed as being fit for the Navy's purpose. Pley immediately asked how much of the Coker Canvas would be required by the Navy, but the delay had meant that a Bristol merchant called Bosell had already contracted most of the local sailcloth makers to meet his client's demand, leaving Pley with just the smaller Coker makers to deliver 50 bolts of canvas for the Navy per month. He cleverly persuaded this group to concentrate on making the 'best sort' of canvas – quality over quantity. The Navy was also slow to pay its bills, however, and as a result Pley had to subsidise his suppliers and the increase in bucking to deliver the quality needed added to the costs of production.

By 1658 George Pley was in partnership with his wife Constance and their son George Jr and Bullen Reymes. They encouraged Constance to continue her intense lobbying of the Navy by writing copious letters to those in influence to support the local sailcloth industry, but despite their combined efforts the local industry went into decline. Nevertheless, the Coker Canvas pattern of production remained the method of choice for the Royal Navy, but the original products were undermined by widespread corruption in the provision of canvas and the extensive imitation of the Coker Canvas pattern by some manufacturers, who kept costs down by reducing the level of bucking and cleansing.

By the close of the 17th century, warships were self-contained communities, capable of operating away from land for months at a time. Dampier's and Cook's first voyages of exploration lasted almost three years, much of it spent in the uncharted waters of the South Pacific. In their cavernous holds, warships carried their own food, fuel, water and clothing, along with all the material and skilled craftsmen they needed to maintain and repair the fabric of their wooden ships. One of those skilled craftsmen was the sailmaker.

Top: The Coker Canvas foretopsail from HMS *Victory*. This sail was last used on HMS *Victory* at the Battle of Trafalgar (21 October 1805). It was pierced 90 times by enemy shot. This photo was taken in 1968.

Above: The Sparkes & Gidley token, 1797.

During the 18th century, Royal Navy ships were not always issued with a complete set of sails. Instead, they were supplied with a sailmaker. He might work alone, if the ship was small, or he might head up a team of people – a sailmaker's mate and two sailmaker's crew. These were busy men. It has been estimated that to make a single topsail for a ship of the line would have taken over a thousand person hours. A 74-gun ship needed several acres of sail, with each sail divided into over 30 individual pieces of canvas. Each sail had to be made by hand, to the exact dimensions of the yard and mast to which it would be attached. Additional sets of sails were required for different weather conditions, such as small but immensely strong storm canvas and studding sails that could be extended out on each side of the ship when the wind was light. The largest sails, the courses and topsails, could weigh over a ton, and even more when wet, which they frequently were.

In order to manufacture sails, the sailmaker was issued with bolts of canvas. These were long strips of material, each 39 yards long and about two feet wide. The material itself was graded according to its thickness and weave, number one canvas being the strongest and heaviest and number six the lightest. The individual lengths were stitched together, edge to edge, to produce larger areas. This method of sailmaking had the advantage of making repairs to damaged sails easier. If a rip were found, the damaged bolt could be cut out and a replacement piece of canvas added.

Threats to national security

By the end of the 17th century a reliance on imported sailcloth was again seen as a threat to national security. Further Acts of Parliament encouraged the growing of hemp and flax in Ireland, free of taxation for imports to England, supporting the revitalisation of the domestic manufacture of sailcloth and improving its quality. An extra duty was placed on imported sailcloth and a bounty added to homemade sailcloth. This culminated in the Act of Regulation of 1736, confirming the taxation intentions and specifying the method of production of a bolt of canvas.

Above: Yarn drying on poles at Coker Sailcloth Works, c. 1900.

The exception to the mildew-prone, grey-coloured canvas being produced across the country was soon recognised as being Coker Canvas.

The craft of making sailcloth had continued to mature in the Cokers such that by the late 18th century it was clear that much of the hemp and flax grown in the villages, as well as the rope and twine, was used for this purpose. Roadside yard spinning of rope and twine was a common sight in the villages, with a young boy or girl turning the spinning wheel and a man walking backwards feeding out the fibres of dressed hemp from the bundle round his waist. In many cottages women would be spinning the finer yarn for sailcloth for the men to use on the hand looms.

However, technological advancements began to impact the industry. In 1733, the Lancashire yeoman farmer John Kay received a patent for his most revolutionary device: a 'wheeled shuttle' for the hand loom. It accelerated weaving by allowing the shuttle carrying the weft to be passed through the warp threads faster and over a greater width of cloth. Designed for the broad loom, for which it saved labour over the traditional process, it needed only one operator per loom (before Kay's improvements a second worker was needed to catch the shuttle). Then along came James Hargreaves' invention of the eight-spindle 'spinning jenny', which around the 1770s replaced the spinning wheel and significantly increased the amount of yarn that could be spun by one person. From this time the increase in flax yarn came about by the factory spinning mills. Virtually all flax for sailcloth was machine spun, allowing the numbers of weavers, and hence sailcloth production, to increase. The production of individual weavers probably did not change very much. Hemp fibres, however, were still spun by hand and remained so until around 1860. In the Cokers, this meant that domestic production continued as the factories developed.

Hillside white with thread and sailcloth bleaching in the sunshine

Throughout this period, there are numerous mentions of the extent of local people involved in making sailcloth. For example, in 1746 John Barratt of East Coker is registered as a sackmaker who takes on apprentices to make sailcloth. In 1770, a reference to John Mullins of West Coker securing a mortgage on a small property, appropriately called 'Hempland', registers him as a sailcloth maker. The release of the mortgage two years later was to John Rendall, who is noted as a flax-presser. Then, in the 1782/1783 marriage register for West Coker, one bridegroom, Onesiphorus Mullins, is recorded as a sailcloth weaver and another, William Graham, as a sailcloth maker. In 1732, Le Herne Cottage in Burton Barton in East Coker is registered as the home of William Taylor, sailcloth manufacturer.

According to reports, noted by Nathan in the *Annals of West Coker*, at this time 'a tourist travelling to the south-west from Yeovil in this period describes his journey … On leaving Yeovil … we were conducted through close sandy lanes to the little village of West-Coker and passed along an extensive valley, bounded by high hills. The country here is full of manufactories and we saw great quantities of thread bleaching in the meadows and orchards, by the side of every rivulet.'

Importantly, in 1774 we have the first record of all British-made sailcloth for the Royal Navy being made in Britain. Despite this, for the next 30 years the Royal Navy did not use Coker Canvas, despite its use as the sailcloth of choice on revenue cutters, packet boats and local smacks. Ironically, it was also the canvas of choice for American men-of-war ships that were fighting against the British Navy. As a result, production of sailcloth decreased in the Dorset and Somerset villages. By the end of the decade, the area was supplying less than 5% of the sailcloth used by the Navy (9,200 bolts – a bolt being a 39-yard roll of canvas) whilst the county of Lancashire produced 60%.

Although records of the supply of sailcloth and canvas from the Cokers in the years of the French Revolutionary Wars (1793–1815) have not been traced, there is a reference in an Admiralty document of 1810 referencing Coker Canvas, thus indicating a familiar use of this term, but with it came more accusations of a campaign against the use of Coker Canvas by officials within the Royal Navy.

Creating work for many

The local industry appeared determined to flourish. In 1780 John Bullock is noted as operating a sailcloth business in East Coker, providing work 'for many'. Bubs Pool House, with its own bucking house, in East Coker is recorded as being the home of a sailcloth maker. In 1815 the house was home to John Cox, a sack and sailcloth manufacturer. In 1826 Bubs Pool was sold to Isaac and James Sant, sailcloth manufacturers. Just four years later it passed to Richard Genge, also a sailcloth maker. The Bullock family, registered as sailmakers, acquired the North Transept of East Coker Church for private prayer. Four sailcloth workers were listed in West Coker. The following landowners in West Coker listed as growing flax for canvas: Portman, Warry, Davey, Daniel, Lond and Baker. Richard Hayward I (1769–1852) from West Chinnock (four miles west of West Coker) is registered as a canvas maker. Over the coming years Hayward and his sons would become key figures in the story of Coker Canvas.

In his *Agricultural Survey of Somerset* of 1797, John Billingsley wrote of the local area: 'There are considerable manufactures of narrow cloth, from four to seven shillings per yard; the quality of which, both for appearance and duration, is not surpassed in the kingdom. ... There are also many manufactures of coarse linen, such as dowlas, tick, &c. also of gloves, girt web, &c. all of which give animation, wealth, and comfort, to the inhabitants of this rich and delightful region.'

The demand for yarn dramatically increased as both the merchant and Navy fleets grew in numbers. With the fresh demand, and despite a bounty being introduced to boost local flax and hemp production, supply could not keep pace. This led to the merchants and manufacturers turning to developing import trade with Russia and northern European and Baltic countries such that in 1796 over 9,100 tons of flax was imported from Russia. Locally, over 100 lead seals from Russian bales of flax dating from this period, including from the city of Archangel (Arkhangelsk), have been found in fields and gardens around West Coker.

Above: John Bullock's seal.

The village of origin

At the start of the 19th century, sailcloth production for the Navy was now becoming concentrated in Scotland. Price was clearly an issue, with Coker Canvas being significantly more expensive per yard than other patterns. In response, local manufacturers upped the ante and the first newspaper adverts mentioning Coker Canvas were published. They make specific mention of East Coker, suggesting that, according to the industrial historian Richard Sims, 'this village is the origin of the style of canvas.' At the same time there were nine sailcloth manufacturers in West Coker.

Following a Commissioners of Naval Inquiry report into misappropriation of public funds in 1802, a further complaint was registered against Viscount Melville (Henry Dundas) in 1806. One of the most powerful political figures in Scotland and First Lord of the Admiralty, he was impeached for the wrongful appropriation of Admiralty orders, favouring factories within his own constituency of Edinburgh rather than those of the west country.

By the end of the year, it was reported that Coker Canvas of several qualities was wanted at Chatham, Sheerness and Portsmouth dockyards in considerable quantities. Within the next few months, it was agreed that the Navy Board order of '5,000 bolts of canvas would be of the Coker kind'. Richard Hayward of West Chinnock, who had supplied the Navy with 400 bolts of sailcloth made in Coker Canvas method, was soon getting repeat orders.

Dundonald hails the benefits of Coker Canvas

In 1806, Lord Dundonald published his review of Naval sailcloth, highlighting that whilst two yards of Navy canvas cost 40d and one yard of Coker Canvas cost 23½d, the latter would last twice as long, resulting in a saving of 42% over two voyages. Dundonald's endorsement saw the rise to prominence of the Coker sailmakers: George Bullock and Messrs Plowman of East Coker and Richard Hayward and Joseph Rendell of West Coker. Following this episode, in 1809 Samuel Hood, sailcloth manufacturer and a distant relation of the Admiral Hood, was entrusted to oversee the purchase and inspection of canvas for the Royal Navy.

On 26 January 1810, sailcloth manufacture in the Cokers was again in full swing, Hood reporting that 'a sufficient quantity of what is called Coker Canvas, can be purchased from different manufactures in the following places, viz. East Coker, West Coker, Odcombe, Halstock, Haselbury and Crewkerne. I have no hesitation that you can get as much of this canvas as you might be pleased to order.' As Crewkerne began to establish itself as the regional centre for the textile industry, Dummett, Perham and Cole opened the area's first mechanised spinning factory in 1807. Sixty years later, the factory was to be taken over by Hayward and renamed the Coker Sailcloth Works.

In London at the beginning of the 19th century, the price of Coker Canvas varied from 2s 11d for first quality to 2s 5½d for sixth grade. The Navy Board then approved Samuel Hood to purchase '605 bolts of Coker Canvas wanted for ten cutters from manufacturers upon the best of terms as well as purchasing a quantity of 24in. East Coker Canvas.' Hood was known to favour the superior quality Coker Canvas, the double thread flax canvas, which was suitable for mainsails, foresails and jibs. To secure this quality canvas he argued for more time, writing to the Navy Board in March 1812 that 'I have no fear of surpassing all other canvas, for every description of vessels, provided Warrants are granted, or proper time given to procure the best Bleached Flax Canvas.'

A further push for the use of Coker Canvas came in 1812 when it was realised that US Navy warships sailing under Coker Canvas had outmanoeuvred the British fleet during one of the many naval engagements in the war with America. Indeed, in the minutes of the Navy Board of January–June 1813 it is recorded that 'the best kind of Canvas (now used in the Navy) is made entirely of Flax and is known by the name of Coker Canvas. It resembles Holland Duck as the best canvas in the world.'

Indeed, in April 1813, Admiral Hood ordered 550 bolts of Coker Canvas from Messrs Bullock and Murley of East Coker and 344 pieces from Mr Joseph Rendall of West Coker for the Navy. Hood reported to the Navy Commissioners, 'I can only add that they are esteemed excellent manufacturing', and a month later Joseph Rendall offered a further 629 bolts. Further testimony to Coker Canvas in contained in a report of September 1813, on the mainsail of the sail ship *Lord Melville* (later to become a convict ship transporting 101 women to

Australia): 'I beg to say, that, I this day saw the *Lord Melville* mainsail. It is in good condition and appears to possess all the advantages and good qualities belonging to Canvas. Well boiled in Alkalis and made of Flax. The Master of the *Lord Melville* has always used Coker Canvas.'

Coker Canvas named Naval standard

Within 12 months, all Royal Navy sailcloth had to be of Coker Canvas pattern (irrespective of where it was made), as it became the 'Navy standard'. This pattern for the Navy's sailcloth was prescribed based on the centuries-old pattern of Coker Canvas. This led to the statement that 'all Navy ships sailed under Coker Canvas', even though most was manufactured in Scotland and elsewhere but always produced to the Coker Canvas pattern.

This edict boosted trade in the Cokers. By the early 19th century canvas manufacturing was the largest and staple activity in the villages, although as a centre of considerable industry making yarn and gloves these industries still employed more in the manufacture of canvas. Nevertheless, sailcloth manufacturing continued to grow. William Arnold is recorded as owning a yarn barton and rope walk on Long Furlong Lane in East Coker, later to become a sack and sailcloth manufacturer whilst living at Redlands Farm, which was the location of a bucking house and a weaving workshop. Edward Taylor (Snr.) of Skinner's Hill Farm, also on Long Furlong Lane and located on the Coker Water, is recorded as being a sailmaker with a yarn barton and a bucking house. At this time, Le Herne Cottage changed ownership to another sailmaker – Samuel Geard. Richard Hayward expanded his production by buying an old water mill in the nearby village of Merriott, converting it from flour milling to flax spinning.

However, things were not all plain sailing. By 1821 there is the last mention of Coker manufacturers supplying the Navy directly with Coker Canvas as a result of intense competition resulting from the Navy's dictate that Coker Canvas was a pattern of choice that could be made anywhere – expansion for some but not for others. There was a general air of concern at the area's

loss of its premium status as the source of Coker Canvas and its impact on the wider community. In October 1828, a meeting of 27 local businessmen took place in the Mermaid Hotel, Yeovil, to agree a minimum fall in the price of cloth to stave off bankruptcy for the local growers, weavers and producers.

Despite this, Coker producers persisted and retained direct sales to the merchant fleets. At this time, Richard Hayward's son even opened a store and warehouse for Coker Canvas in London, and five sailcloth makers were still registered in West Coker. Then, in 1825, the company Richard Hayward and Sons was established when three of the founder's sons became partners in the business, allowing Hayward (Snr.) to take a back seat in the day-to-day running of the company.

Similarly, expansion was on the agenda, especially in West Coker, and for the Moore family at The Wash, the Baker family of Manor Street and the Rendell family of the Cross.

Above: Former weaver's cottage, Back Lane, East Coker.

The Rendell family played a key role throughout this period of success. The family was regarded as being sturdy and strong characters in the 16th century, as Nathan writes, 'they may have been poor but there was no lack of vitality about the Rendalls [sic] leaders in the industry locally'. Indeed, a John Rendell is noted as a flax-dresser in 1772, at a time when looms in East and West Coker became busy with making sailcloth. Joseph's son and daughter, Israel I (1778–1827) and Harriett (1788–1849), became ever-important figures in this story. For Harriett this was through her marriage to Edward Taylor, a sixth-generation sailcloth manufacturer who, together with his father, built the sailcloth factory in East Coker, while Israel II was to become the largest sailcloth manufacturer in West Coker.

The Rendells had been involved in sailmaking since Lazarus Rendell left the use of some equipment to his nephew Joseph, who went into partnership with his sons John and Israel Rendall I (b. 1778). Joseph Rendall was an established hemp and flax trader and sailcloth manufacturer of East Street, West Coker who supplied canvas to the Navy. The Rendell's partnership traded as Joseph Rendell and Sons until 1801, when Israel I and Joseph decided to trade separately. Israel bought land below The Wash on High

Street (formerly Duck Lane) in West Coker from the Portman family to set up his own sailcloth works, building Millbrook House next to the works as his family home. Israel I died in 1841 and the business passed to his son, Israel II, who focused on twine manufacture. It was Israel II who by 1880 had created 'a very desirable residence with six bedrooms fitted with a bath and w.c. ... most pleasantly situated at the west end of the village ... and occupied by mr. W. J. Dawe', who was employed by Rendell to manage his new twineworks at a time when factory-based power looms took over from sailcloth production.

Whilst factories such as Israel Rendell's were responsible for the bulk of the sailcloth production, there is evidence that it also remained a cottage industry. There were often four or five looms in many small cottages, such as those in Duck Lane (West Coker) and Ten Houses and Cross Cottages (East Coker). Eight weaver's cottages are recorded on Back Lane, East Coker, each with hand looms. It was in one of these cottages that Mrs Mary Rendall (d. 1929) claimed to have woven the sailcloth for a yacht owned by Queen Victoria and to have been paid a rate of 5s or 5s 6d per piece.

Continuity, investment and recovery

Back in North Coker, the imposing Devonshire Cottage was first occupied by Samuel Hutchings sailcloth manufacturers and then passed to Edward Taylor, also a sailcloth manufacturer. This should not be a surprise, as the cottage had a yard, a yarn barton and a flax house on site. Later, a new Devonshire Cottage on the same site became the family home of Felix Drake, who owned East Coker Mill and Rope Walk in Halves Lane, East Coker. In West Coker, George Gould II (1808–1873) initially rented a twine walk in West Coker in 1873, which his son, Job Gould Snr. (1843–1915) later established as the West of England Twineworks at The Cross. Tension between families did exist, indeed, Job is reputed to have said 'there was a great deal of jealousy and persecution [towards his father] on the part of the other twine firm in the village' (owned by Israel Rendell).

Considerable investment in the local transport infrastructure was taking place to help boost business. The opening of the Beaminster Tunnel improved road access by connecting the Cokers with Beaminster and

Left: Devonshire cottage, c. 1938.
Top: Taylor, later Drake's factory, c. 1880.
Below: Sailcloth power looms at the Coker works of R. Hayward & Co.

The Making of a Global Brand

Above: Advert for Drake's webbing factory, 1920. Originally built as a power loom sailcloth factory for Edward Taylor, Felix Drake converted it for webbing, c. 1879.

the port of Bridport to the south, while the River Parrett navigation system providing access from Langport to Bridgwater and the Bristol Channel was also improved. In 1835, improvements locally were undertaken by the Yeovil Turnpike Trust, near the public house known today as the Quicksilver Mail at the top of Hendford Hill. Originally known as the Phoenix Stagecoach, the Quicksilver Mail became the official mail service between Exeter and London and used this former coaching inn as an overnight stop before the 11-hour journey to London.

In 1846, a special Commissioner's report on the future of Yeovil Market suggested a number of improvements, and in April 1848 the *Sherborne, Dorchester and Taunton Journal* reported on the 'opening of The New Market House in South Street, Yeovil. Over the dry goods market is a large Flax Room, which on occasions was pretty well filled with the commodity for which it was designed.' In his 1850 book *A Sketch of the Town of Yeovil*, Daniel Vickery (later to become the editor of the *Yeovil Times* in 1856) observed that 'The Market Day (in Yeovil) is every Friday, every alternate Friday is The Great Market. ... Hemp and flax are sold in great quantities.'

In 1853, the first broad-gauge railway arrived in Yeovil, connecting Taunton to Yeovil Hendford station, constructed by the Bristol and Exeter Railway Company. The Wilts, Somerset and Weymouth Railway Company opened Pen Mill Station in 1856, and a year later the line was extended to Weymouth and its harbour. In June 1860 the Salisbury and Yeovil Railway Company opened Yeovil Hendford station of the Bristol and Exeter Railway, and in 1861 Yeovil Town station opened. This was soon followed by the opening of the Yeovil to Exeter railway in July, with both Yeovil Junction and Sutton Bingham stations serving East Coker. The rail link to London reduced the journey time to just four and a half hours.

Such was the recovery of business that in 1851 it was estimated that 9%–10% of the population of the two villages were employed in sailcloth weaving. However, a major shift in production was taking place as the processes became more mechanised and home working declined. The number of weavers using hand looms in Back Lane, for example, had declined from eight to just one following the establishment of power loom weaving mills in East and West Coker.

It was at this time that Edward Taylor Jnr. sold his existing textile interests to fund and develop a new steam-powered sailcloth factory in Halves Lane, East Coker and by 1870 he was employing 185 local men there. Simultaneously, James Wall Row expanded the Coker Sailcloth Works in Crewkerne by eight times its original size, thus helping to boost sailcloth production from the whole area once again.

Above: North Coker House.

A remarkable twist of fate

The winds were turning once again, and the 1860s witnessed yet another blow to the local sailcloth industry. This time it was not competition from other parts of Britain that caused the decline but rather the emergence of steam power taking over from sail on sea-going vessels. This directly and immediately led to a dramatic decline in demand. The local industry became more centralised in Richard Hayward's newly acquired Coker Sailcloth Works in Crewkerne, which he gradually expanded, improved and

Above: Brunel's steamship, the SS *Great Western*, in choppy seas.

modernised over the next 30 years to be more competitive as cotton increasingly became the textile of choice for sails across the world. Locally, over 300 men and women were laid off in the Cokers and, despite a short-term recovery, the traditional sailcloth industry in the villages effectively ceased in 1870. Along with it, the manufacturing of hemp sacking, another speciality in the Cokers, came to an end. In 1851 there were 25 registered sack and bag manufacturers in the villages. Though this had risen to 35 by 1861, by 1881 just 2 remained.

There followed a gloriously ironic twist of fate in the dramatic decline of sailcloth manufacture in the Cokers. In 1865, the Bullock family, wealthy landowners with properties in Somerset and Wiltshire, built North Coker House as their main residence and invested in other improvements for East and North Coker. Over the next hundred years the Bullock's would adopt a series of surname changes: Troyte-Bullock, Troyte-Chafin, Bullock-Grove. In 1920, East Coker's two large local landowners, the Helyar and Troyte-Bullock families, started the disposal of their lands.

The 1920 auction of the Troyte-Bullock Estate included their sailcloth factory in East Coker making it clear that he would evict any man from one of his cottages for taking employment at his former factory in East Coker, now owned by Drake's, during slack time on the land. The Grade II listed house was eventually bought by Cyril Maudslay (1875–1962) for himself and his wife, Dorothy (1892–1977), together with some of the estate's other properties, including Chantry Cottage and the East Coker Village Hall.

As fate would have it, Cyril's family owned the Maudslay Engineering Company of Coventry, whose steam-powered 'Maudslay Marine Engine', developed in the 1830s, had been fitted in London by the then Maudslay, Son & Field Engineering Company. These two steam engines provided 750 horsepower, enabling the SS *Great Western* to embark on its maiden voyage and take just 15 days outbound from Bristol to New York in April 1838 and 14 days for the return journey.

Designed by Isambard Kingdom Brunel and built in Bristol between 1836–37, the SS *Great Western* was the first purpose-built paddle passenger steamship for crossing the Atlantic and the largest of its kind in the world. With a crew of 60, it accommodated 128 first-class passengers and 20 of their servants. She made a total of 74 Atlantic crossings and was still making Blue Riband crossings in 1843 before being used as a troop carrier during the Crimean War.

In addition to the two Maudslay Marine engines, SS *Great Western* was designed with four masts whose sails provided auxiliary power and helped keep the ship in balance during rough seas, helping kickstart the demise of sail to steam power on ocean-going ships. Ironically and cruelly, the sails used on the SS *Great Western* – a ship fitted with steam engines made by the family who were to own North Coker House, the home of John Bullock, the greatest sailmaker in East Coker – were made from Coker Canvas.

Top: Job Gould's West of England Twineworks, c. 1890 (John Albert Gould is in the centre of the back row).
Above: Drake's office staff, c. 1960.

Above: Reels of twine at Dawe's Twineworks.

A new era of success

In the Cokers, two important things happened with the demise of the manufacture of their famous traditional sailcloth destined for the Navy. Firstly, new products had to be made, meaning that innovation became the order of the day. Secondly, Coker manufacturers diversified further as they turned again to twine and rope and added webbing to their portfolio of activities.

By 1872 Felix Drake (as F. Drake and Co.) was now the owner of Devonshire Cottage and converted Edward Taylor's sailcloth factory into a twineworks and webbing weaving factory in Halves Lane, East Coker – a business that survived until 1987. The scale of the business was reflected in their letterhead, which boasted warehouses in Dublin, Norwich, Birmingham, London and Manchester. At the same time, Richard Henry Hayward (grandson of Richard Hayward I) left Harrow School at age 18 to join the family's tail twine and weaving mill near Crewkerne. It was about this time that George Gould moved the business to Martock, setting up at the Parrett Works. After his death, Job brought the business back to East Street in West Coker (the site where the buildings of his West of England Twineworks still remain to this day).

At this time, Israel Rendell, the owner of the Millbrook Twineworks in West Coker, decided to retire to Bournemouth, where he died intestate. The twineworks at Millbrook House was bought by William John Dawe, who had worked as Rendell's manager, clerk and travelling salesman, and it was Dawe who decided to concentrate on twine making at the Millbrook site using the existing facilities until he built his own new works developed by Francis John Dawe (William's youngest son).

The new buildings were completed in 1899 when William Sibley & Son, of the Parrett Works, Martock, built and installed the machinery for the sum of £990 12s. The 'mechanised' process of twine manufacture needed more cover than the traditional ways but still required airy conditions, hence the open-sided 100-yard walk, which would also have allowed work to continue during inclement weather. The first edition of the relevant Ordnance Survey map of 1887 shows a narrow walk to the north of Millbrook House but nothing on the present site. By the time the map was revised in 1903 the twineworks buildings had been completed and are shown as substantially as they are today.

Above: Henry Dawe, painted by British artist Frederick Brueton, 1892.
Right: Robert Harvey with a new wagon he built for Dawe's Twineworks, c. 1900.

Dawe and the Millbrook Works

The Millbrook Works was constructed entirely of timber, bolted together, and roofed with double Roman clay tiles made by A. G. Pitts of Highbridge. The east end of the walk, including the office and drying area, was enclosed by walls of light boarding, while the west end consisted of the old stone boundary wall. The loft, or twisting walks, was only two-thirds boarded over, with open spaces on each side, further increasing the airiness of the building. The only solid brick structure was the engine shed, where, judging by the size of the coal store and the quantity of coal ordered in the early years, there must have been a steam engine. There was then an oil-gas engine, of which the concrete base and holding-down bolts survive, as does the large exhaust pipe, which was led up through the drying room, where it must have provided considerable heat. Later still, electric motors were installed as well as homemade bar heaters in the drying room. It was here that Dawe pioneered a method of making the twine shine by applying a polish of tallow.

In January 1899, the firm's annual income was £88 13s 2d with expenditure of £83 12s 4d. In December 1914, the annual income was £91 15s with expenditure of £85 14s 9d. In 1914 the business had a healthy bank balance of £3,034 17s 8d. In 1910 the firm purchased materials to the value of £3,258 15s 10d, selling goods to the value of £5,227 19s 8d. Also in 1910, advertising space of a tenth of a page was taken for 12 months in *The Cabinet Maker* and *Complete House Furnisher*.

Dawe had dealings with firms as far afield as The Kirkcaldy Spinning Co. in Fife, Scotland and Max Muller & Co. of Ghent, Belgium. Local firms dealt with were: Baiston & Co., Poole; Burton Spinning Mill Co., Bridport; Drake & Co., East Coker; Hancock & Co., Taunton; Hayward & Co., Coker Works, Crewkerne; Rawlings & Sons, Frome; Smith & Co., The Parrett Works, Martock and Tucker & Co., Slape Mill, Bridport. W. S. Dawe eventually took over from his father and was succeeded by Frank Dawe, who died in 1937. His son, Henry Stuart Dawe, ran the works until its closure in 1968.

Stuart, an Oxford graduate and pilot in the Royal Flying Corps, had little background knowledge of twine making, yet persisted with the family business. Even into the late 1930s, Dawe's was employing about 30 villagers – 20 men in making twine and 4 women working in a separate building balling and preparing the finished products for dispatch to clients.

Below: Drake's Webbing Factory, 1914. (Felix Drake is standing fourth from the right.)

A Global Export – From Virginia to New South Wales

In 1816, as reported in *The Review of the Mercantile, Trading, and Manufacturing State, Interests, and Capabilities of the Port of Plymouth With Miscellaneous Additions by Other Persons, and Notes* by William Burt, it is clear that Coker Canvas was being exported through this Devon port. *The Review* states: 'the temptation, which some cannot resist, to purchase at very low prices, canvas, made in the north of England, of an inferior description; and thirdly, the preference given, from its elder reputation, to the Dorsetshire and Somersetshire cloths, particularly to that made in the village of East Coker, in the latter county, which is deservedly held in such esteem, that much imposition is daily practised by stamping canvas as its produce, manufactured at places remote from it.'

The reputation of Coker Canvas had a global reach. Recent research by Jonathan Sherriff reveals that the *Norfolk Gazette* and *Publick Ledger* of Monday evening, 17 December 1810, and again on 8 January 1812, J & V Southgate were selling East Coker Canvas from the warehouse on Woodside Wharf, Norfolk, Virginia, whilst in the archives of the National Library of Australia there are some 80-90 items, almost entirely newspaper and journal advertisements for Coker Canvas (some specifically mentioning East Coker) dating from 1829–1859, offering varying amounts for sale and in auctions. Most of these adverts appeared in the Sydney and New South Wales media. The exact use of Coker Canvas in Sydney is still to be researched, however, the fact that most of the adverts for Coker Canvas appear on the front pages of these Australian newspapers suggests that it was a product of some importance.

One such advert, from about 1840, is for the sale of 30 bolts of East Coker Canvas by Messrs. Harper Blundell & Co., whose store was located on Bligh Street at the heart of the Rocks area of Sydney Harbour. Sadly, in 1846 the company petitioned for insolvency, with debts of over £8,000. Close by, on the corner of George and Hunter Street, also in 1846, in the *Sydney Morning News* of 30 September the publican and auctioneer Phillip Cohen was advertising the sale of bleached East Coker Canvas.

Coker Canvas was well represented in the *Official Catalogue of the Great Exhibition of All Nations 1851* and in the *British Trade Journal* of 1856. Throughout the early years of the 20th century *The List of Manufacturers of British Goods Exported to New South Wales* had Hayward's as a recognised exporter. Even as late as November 1936, the *Pacific Islands Monthly* was carrying adverts by W. Kopsen (Ships Chandlers), of Clarence Street, Sydney, for Hayward's Coker Canvas.

Above: Coker Canvas adverts in the *Sydney Morning News*, 1846 and *Pacific Islands Monthly*, 1936. National Library of Australia.

Giving the Past a Future and the Future a Past: Dawe's Twineworks

'Still and still moving' (from *East Coker* by T. S. Eliot)

Today, Dawe's Twineworks, with its impressive 100-yard covered ropewalk, is England's only surviving and working twineworks. Saved, restored and brought back to life as it has been by its community, its rich history has now to be celebrated.

'I remember West Coker when its time was daily recorded by the factory hooter blowing at 8am, 1pm and 6pm at both twine factories. Life then may have been simpler and more honest and in many ways better than the situation we have today.' (Interview with a resident of West Coker by David Shorey in *The Book of West Coker*, 2008.)

After its closure in 1968, the Dawe's Twineworks lay vacant until 1980, when an Industrial Museum for Somerset was still a possibility. The Somerset Industrial Archaeological Society approached Mr Dawe, asking if he would donate the redundant machinery for future display in the new museum, to which Mr Dawe agreed, and in December 1980 a full measured drawing and photographic survey of the works was undertaken ahead of the planned removal of the machinery. However, with his wife extremely ill, Mr Dawe asked the Society to delay the removal of the machinery. Sadly, Mrs Dawe passed away, and shortly after her death Mr Dawe also died.

Before the Society became aware of what had happened, the property was sold to a boat and general dealer and the Works became full of his stock of spare parts. At this point South Somerset District Council became interested in securing the future of the Works, and the buildings were listed Grade II in 1984.

The collective effort

In 1996 the Council became actively involved in seeking to secure the future of the Twineworks and convened a meeting of interested parties, including the Royal Commission on the Historic Monuments of England, the Industrial Buildings Preservation Trust, the Royal Dockyard at Chatham, Somerset Industrial Archaeological Society. With the owner's consent, South Somerset District Council asked Somerset Industrial Archaeological Society to help list and tag all the machinery and ancillary equipment. This work was completed in January 1997.

Although there were other surviving rural rope and twine walks throughout the country, Dawe's was the only one to have retained all of its machinery. This prompted a full investigation of the site by the Royal Commission on the Historical Monuments of England (which merged with English Heritage in 1999). The results were published in 1998 and the listing of the Works was upgraded to Grade II* in 1999. At the same time, the Industrial Buildings Preservation Trust commissioned a feasibility study to assess the possible future use of the Works and the cost of renovation of the buildings and machinery, also completed in 1999. Two years later, South Somerset District Council decided to acquire the Works and then spent several years negotiating with the owner, without success, eventually compulsorily purchasing the property in 2005 in order to save it, promote its restoration and secure its future.

A community rewarded

Coker Rope and Sail Charities, set up specifically by the community to manage the conserved property, with help from South Somerset District Council, English Heritage, Somerset County Council, the Architectural Heritage 'Challenge Fund', the Headley Trust, the National Heritage Lottery Fund, the Andrew Lloyd Webber Foundation and the Carpenter's Fellowship, undertook essential stabilisation and repair works in 2010. However, Dawe's Twineworks remained in a parlous condition and was on English Heritage's 'Listed Buildings at Risk Register, category 'A' – 'Immediate risk of further rapid deterioration or loss of fabric; no solution' – for some considerable time. The whole Works, particularly the existing twine-making machinery and equipment, which is the most important part of the whole project, was left extremely vulnerable and in need of conservation.

Top: Dawe's ropewalks in West Coker, showing the finishing walk nearest the camera and the twisting walk near the trees, c. 1915.
Above: Twinewalkers at Drake's Factory, c. 1930. Left to right: Walter Stevens, F. Rendell, E. Purchase, T. Thomas and T. Penny.

However, there has been great progress since 2013. The engine house, office, store and east end of the walk have been repaired and restored. An Arts Council 'PRISM' grant then funded the restoration of much of the twine-making machinery and a 'new' oil engine to power it. In spring 2015, outer rows of oak posts were also placed to support the twine walk.

In September 2016, the National Heritage Lottery Fund approved the second and final stage of a CRST application for a grant to complete the restoration of the Twineworks and to provide visitor facilities. The project proceeded very well and the work on the old buildings and machinery is now complete. The scaffolding was at last removed in 2017, and the twine-making processes are now set up along the full length of the 'walk'.

The covered twine walk is 90m long, with the upper floor housing the twisting machinery, where strands of yarn are twisted by the wheels, driven by belts linked by shafting to the oil engine in the adjacent shed. The twine is made

Left: Dawe's Twineworks after restoration, 2019.
Above: Dawe's Twineworks before the restoration.

Above: The finishing walk at Dawe's Twineworks following the restoration.

into 180m loops and placed on rotating pulleys at each end of the walk. The moving twine is washed in water baths before being sized. Twine was normally polished, but not that destined to be used for sailmaking.

The new Visitor Centre, complete with exhibitions and a café, has replaced a 1930s bungalow. The site is now open to visitors and welcomes educational groups and local societies. It is the venue for themed dinners, tours and talks, and also hosts concerts and the regular, highly innovative Od Arts Festival by Simon Lee Dicker and the OSR Projects team, based in West Coker. Today the Dawe's Twineworks is full of life, living history and sensual delights, complete with its own cider mill producing a light morning cider.

Elsewhere, in West Coker, the main structures of Job Gould's original West of England Twineworks are still to be seen and until recently it was active producing 'Coker Cordage' (see supporters Peter and John Gould).

Above: Volunteers at Dawe's Twineworks.
Right: Dawe's Twineworks after restoration.

Giving the Past a Future and the Future a Past: Dawe's Twineworks

Above: Spools of twine at Dawe's.
Right: Flax bundles with lead seals.

JOB GOULD & SON LIMITED

Directors: ALBERT GEORGE GOULD, JOB ALBERT GOULD.

West of England Twine Works

WEST COKER YEOVIL SOMERSET

Telephone: WEST COKER 327. Telegrams: JOB GOULD, WEST COKER.

OUR SPECIALITIES

Mattress Twines and Laid Cord

Mattress Tufting Loops

Upholstery Twine and Spring Cords
for Furniture, Motor Cars and Theatre Seats

Stool and Chair Seating Cords

Brush Makers' Twines and Cords

Packing, Parcelling and Baling Twine

Armature Twine for the Electrical Industry

Bat Binding Twines

Page Cord for Printing

MADE FROM FLAX AND HEMP

Working at Dawe's

Here Reg Warr describes his experiences in an interview recorded in Shorey, D. with Dodge, M. and Dodge, N. (2008) for *The Book of West Coker A Pictorial and Social History of a Somerset Village and its People,* published by Halsgrove:

'Twine making was a normal part of my early life. My father worked full-time at the Millbrook Twineworks then owned by Frank Dawe. We lived in a tied cottage owned by Mr. Dawe. The five-roomed cottage (on Millbrook Street) had been a store house for sailcloth 250 years earlier. If any of us displeased Mr. Dawe we could be turned out at one week's notice from our cottage. Workers were not allowed annual paid holidays at the Twineworks until 1938.

'The strongest and best twine was made from locally grown flax and hemp. Women and children did the pulling of flax. My mother, when she was a child, used to help Grandma. 'It used to make my hands so sore,' mother said.

'I reckoned that I would walk an average of 45 miles a week in the ropewalk. I loved the warm oily smell in the engine room with the large flywheel and the leather

---

Above: Advert for Job Gould & Son, c. late 19th century.
Top right: Jim Baker retired from Drake's factory aged 85, having worked there since he was ten, 1947.

belt going round driving the power onto the connected pulleys and shafts. It was so noisy; I could not hear myself speak above the clanking of the beautiful green shiny Hornby engine. It was much quieter in the twisting loft. The flax and hemp made its own comforting, almost musty, smell but down on the ground floor the dominant smell was from the wet size that was used to stick the fibres on the twine.

'Two brothers, Jack and Tom Stroud, twisted the heavier ropes by hand on the twisting wheels outside in all winds and weathers. They were little men, almost blind and had worked all their years for a pittance until they died. They were so arthritic they could hardly walk straight. They lived together in a tiny cottage. They ate bread and cheese and drank cold tea from a bottle. When my mother offered them cake they refused saying they preferred cheese. No one earned much in the twine trade for their 48-hour week. When I worked there in 1939 I was paid 18s for a 48-hour week'.

Below: Andy Parker's installation at Dawe's, OSR Arts.

Giving the Past a Future and the Future a Past: Dawe's Twineworks

The Sailmakers, HMS *Victory* and the America's Cup

Gold medal for Coker Canvas

In terms of innovation, around the 1880s Richard Hayward and Co. saw the opportunity to specialise in sailcloth for yachts after recognising the emergence of luxury yachts and the growing interest in classic yacht racing. The company also started using cotton as the canvas for yacht sails made in the Coker Sail Cloth Works, Crewkerne.

The Haywards succeeded in establishing a business relationship with specialist yacht sailmakers Ratsey and Lapthorn of the Isle of Wight – the world's pre-eminent sailmakers – who were already involved in the newly established America's Cup. In 1878 Edward Taylor's East Coker sailcloth business was declared bankrupt and ceased trading. Edward Taylor was then recruited by Richard Hayward & Co. to manage their sailcloth factory in Crewkerne and he was to become the key figure in producing the quality canvas demanded by Ratsey & Lapthorn.

Due to the split in the family and confusion over the names of the various divisions within these businesses, Richard Hayward & Co. of the Coker Sailcloth Works in Crewkerne issued a public notice stating that it was this company, not that of his brother, that had won a gold medal for their Coker Canvas at the 1883 International Fisheries Exhibition, South Kensington, London – one of the many science, technology and cultural World Fairs of the Victorian era – and that his company was working at 'full capacity', supplying the Navy with 400 bolts of Coker Canvas a week from his Crewkerne factory.

Richard Sims has recently discovered that Edward Taylor's uncle, George Taylor, had married Mary Jane, the daughter of James Lapthorn – the founder of the Lapthorn sailmaking firm that would later merge with Ratsey. Sims reflects on the fact that, 'given business protocols at the time, it seems like that the couple would have met during a time when James Lapthorn and Edward Taylor were doing business together. This might have given Taylor insight into yachting sailcloth and the reason for Hayward giving him a job.'

Above: Portrait of Lord Nelson by Lemuel Francis Abbott.

The sails for HMS *Victory*

The firm we know today as Ratsey & Lapthorn is descended from two distinct and separate British firms: Ratsey, based in the Isle of Wight, and Lapthorn of Gosport. As businesses, Ratsey and Lapthorn have been making sails for over 225 years. They are international experts in heritage craft of sailmaking for classic yachts. The company is regarded as a venerable institution that has set the world standard for high-quality sails.

The Ratsey sailmaking heritage goes back to George Rogers Ratsey, who set up shop as general traders between Cowes and the mainland. In 1790, George's uncle John arrived with 20 years of experience of working in sail lofts, persuading George to start making sails for local vessels, including those built by a relative – Charles Ratsey.

Above: HMS *Victory*, Portsmouth Historic Dockyard.

Above: The *America* winning the sailing match at Cowes for the Club Cup, open to yachts 'of all causes and nations'.

George soon made a name for himself by making better sails than anyone else. In 1815, for instance, the Ratsey-outfitted yacht *Waterwitch* easily beat *Pantaloon*, a naval vessel of the same size. Admiral Sir Putney Malcolm summoned Ratsey for an audience: 'I want you to tell me what there is in your sails that makes them superior to all the fleet.' As great-great-grandson Ernest once told the story, George Rogers smiled, cleared his throat and said nothing. 'Hell,' says Ernest, 'they didn't think he was going to give away his patterns, did they?' Indeed, it is a company that prides itself on innovation through science and materials that makes faster yachts.

In 1795, Lord Collingwood of the British Admiralty was a strong advocate of what he saw as the stiffer and better sails coming out of Ratsey's loft. This appears to have led to them providing some of the sails for HMS *Victory* that Lord Nelson used in the Battle of Trafalgar in 1805; the topsail remains in the National Museum of the Royal Navy at Portsmouth's Historic Dockyard, complete with cannon ball holes.

Given that Coker Canvas was the sailcloth of choice for the Navy and Lord Nelson and that Ratsey was the preferred sailmaker for the Navy, it is axiomatic that there was a strong trading relationship between the Coker sailcloth makers and Ratsey that was flourishing well before Ratsey and Richard Hayward's Coker Canvas Works partnership of the late 19th century to produce sails for the America's Cup. The sailmakers would have recognised the importance of the good-quality twine that came from the Cokers. The sewing of sailcloth into sails demanded large amounts of twine. For example, it is estimated that the fore topsail of HMS *Victory* would have required 3,500m of twine and the *Victory* had over 50 sails.

It was George's son, another Charles, who took control of the business in 1844 and with his son Thomas began to develop the business. Today, the Ratsey name is the oldest sailmaker in the world in continuous business.

James Lapthorn served his sailmaking apprenticeship in south Devon but in 1815 moved to manage a sailmaking business in London. There he was approached by Captain Lyon of the Royal Yacht Squadron, under whose patronage he set up a sail loft in Gosport in 1825 and within five years was advertising the sale of yachts as well as sails. James encouraged his sons, James II and Edwin, to join the company, but following the death of both his brother and his father in the 1860s, Edward struggled until his own son and a cousin joined the company. Boom times followed.

By 1880, Lapthorn had cornered the market for yacht sails – 75% of all British demand was serviced by the company. At this time Ratsey served 5% of demand. Just two years later, Lapthorn's share had reduced to 50% and Ratsey's had grown to 8%. It was to his credit that Edwin Lapthorn had the presence of mind to seek a merger with Tom Ratsey, so in January of 1882 Ratsey & Lapthorn was born.

Coker Canvas and the America's Cup

Today, Ratsey & Lapthorn is synonymous with the America's Cup, being the sailmakers of choice, providing sails for competing yachts from both sides of the Atlantic for over 100 years. It all began in 1851. George Ratsey, who was one of the founders of the Royal Yacht Squadron in Cowes, helped to establish a regatta to compete for the Hundred Guineas Cup (it was actually called the Cup of One Hundred Sovereigns) – a trophy which later became known as the 'Auld Jug', or the America's Cup. The cup itself is a bottomless silver ewer standing 27 inches high. It was presented to the Royal Yacht Squadron by Lord Anglesey who, as Lord Uxbridge, had been another of the founding members of the Squadron.

Richard Sims argues: 'It was on 22 August 1851 that the US schooner *America* wrote itself into the history books.' Sent by its owning syndicate to represent the New York Yacht Club in the Royal Yacht Squadron's annual regatta in an extraordinary race over 53 miles lasting for over ten hours, *America* won against fifteen contestants to win the prestigious Hundred Guineas Cup. This rocked the yachting establishment, the more so as it had cotton sails rather than the flax used in Britain.' Just two years later, Lapthorn were advertising 'sails on the American principle.'

Six years later, in 1857, the winning syndicate in New York permanently donated the trophy to the New York Yacht Club, renaming the trophy the America's Cup and requiring it to be available for perpetual competition between defending and challenging yachts. This makes the America's Cup (named after the yacht, not after the country) the oldest international sporting trophy in the world. The first competition took place in 1870 and in the years since there have been 36 challenges. The New York Yacht Club held the trophy for an unbroken 132 years before losing it in 1983 to Australia. Since then, the winners have come from the USA, Switzerland and New Zealand.

In August and October 2024, Barcelona will host the 37th Louis Vuitton America's Cup. Three preliminary events took place in late 2023 and early 2024, prior to the Final Preliminary Event and what is known as the Challenger

Selection, with the actual race starting on 12 October 2024. The current defenders are the Emirates Team New Zealand, representing the Royal New Zealand Yacht Club. There are five challengers who have been encamped in Barcelona since summer 2023 to prepare for the competition. They are INEOS Britannia, ALINGHI Red Bull Racing, Luna Rossa Prada Pirelli Team, NYYC American Magic and the Orient Express Racing Team.

Above: Mark Matthews, sailmaker, stitching with Coker Canvas.

Above: Sailmaking tools used at Ratsey & Lapthorn by Mark Matthews.

The first challenger was *Cambria*, in 1870. This schooner was owned by John Ashbury but built by Michael Ratsey at his yard in Cowes and fitted with sails made by Lapthorn in Gosport. *Cambria* came eighth in the race, with the winner being the U.S. defender, *Sappho*. Undeterred, Ashbury issued another challenge for the following year. Commissioning Michael Ratsey to design the yacht *Livonia,* he also commissioned Charles Ratsey to make the sails – this time using American cotton.

After their successful merger in 1882, Ratsey & Lapthorn, in turn succeeded in securing the rights to provide the sails from the Hayward's Coker Canvas sailcloth. In 1887, George Lennox Watson, a specialist yacht designer, built *Thistle* specifically to win the America's Cup with cotton sails made by Ratsey from cloth woven by Richard Hayward in the Coker Sailcloth Factory in Crewkerne to the Coker Canvas style. Hopes were high, but *Thistle* sadly failed to bring the trophy home. Despite this, Hayward's now sold all their yachting sailcloth to Ratsey & Lapthorn, as the contract was extended. In 1893 Richard Hayward & Co. supplied the Coker Canvas to Ratsey & Lapthorn for the Royal Yacht Squadron's challenger vessel, the Earl of Dunraven's *Valkyrie II* in New York City. *Valkyrie II* lost to the American yacht *Vigilant,* but there was widespread praise for the quality of her sails.

In order to ensure the best quality sailcloth, that for each yacht was woven on a single loom which was specifically set up for the yacht's characteristics. Hayward's gave these looms the names of the yachts. It was at this point that Howard Gould, *Vigilant* owner, asked Ratsey to make a suit of sails for him, to which Ratsey replied, 'No, let this be America against Britain in ships, sails and seamen.' This changed after 1901, when Ratsey was persuaded to open a sail loft in New York.

In 1903, following the setting up of Ratsey & Lapthorn's New York loft, Richard Hayward & Co. sailcloth was continued to be used by both competitors in America's Cup in New York – *Reliance* and *Shamrock III*. In 1920 Coker Canvas was again used by both yachts competing in the America's Cup – *Shamrock IV* and *Resolute*.

Coker Canvas continued to be used by both competitors for another 30 years.

Indeed, in the 1937 America's Cup the sails for the challenger yacht, *Endeavour II*, were made of Coker Canvas – it was reported that the mainsail on its own needed 650,000 individual stitches, all done by hand. In this same year, British Pathé News made a film about the revival of sailmaking in Crewkerne due to the surge of interest in luxury yachts and the success of the America's Cup.

The origins of yacht racing in America

Racing yachts was a novel idea for most Americans in the middle of the 19th century, despite the seafaring traditions of the East Coast states. With few decent roads, travel and trade up and down the East Coast was dominated by the sea. In the first three months of 1851 it was estimated that more than 6,500 vessels passed Cuttyhunk Island between Boston and New York. There were some boats built for pleasure and impromptu racing – Boston had become the first yacht club in North America in 1835. It was to be New York, an island and a seaport, where yacht racing became fashionable – and where Ratsey & Lapthorn focused their business.

It was an extraordinary group of brothers named Stevens (no relation) who, more than any other, popularised the sport of yacht sailing in North America, especially in New York. The brothers' father, Colonel John Stevens III (1749–1838) – whose grandfather emigrated to America from London around 1695 – was a lawyer, engineer and inventor who constructed the first U.S. steam locomotive, the country's first steam-powered ferry and the first U.S. commercial ferry service from his estate in Hoboken, New York. He was influential in the creation of U.S. patent law but made his fortune in designing and building railways and, ironically, steamships.

It was John Cox Stevens (1785–1857), the eldest of the Colonel's thirteen children, who inherited the family's business and associated wealth together with the Hoboken Castle and Estate overlooking the Hudson River. Regarded

as a 'bit of a sportsman', John often frequented New York's horse racing tracks, wagering large sums and becoming president of the New York Jockey Club as well as owning the racehorse American Eclipse. He is credited with introducing cricket to America and installed a baseball diamond in a park on the Estate called the Elysian Fields.

However, it was yacht racing that captured his imagination and monopolised his life. Indeed, he is best known for founding and serving as the first Commodore of the New York Yacht Club in 1844 – a post he was to hold for 11 years. He took ownership of his first yacht, *Diver*, in 1809 at the age of 24 but soon progressed to larger vessels, encouraging other rich New Yorkers to build racing yachts and accept his challenges in 'trials of speed' on the Hudson River.

John's brother Edwin Augustus Stevens (1795–1868) inherited his brothers love of sailing and, together with three classmates from Columbia College and his older brother, met in the New York Yacht Club's clubhouse on the Hoboken Estate to establish a syndicate to build a new vessel to race in England and Europe. During a delay in the construction of their yacht *America*, Commodore

Above: *Britannia* and *Ailsa* racing on the River Clyde.

Vigilant and *Valkyrie II*. America's Cup, 13 October 1893.
Oil on canvas, 30" x 40", 2002.

John Stevens received a letter from Lord Wilton inviting the New York Yacht Club to race in the Solent between the Isle of Wight and the English mainland. The challenge was accepted, but not without a lot of help from George Ratsey. Stevens was encouraged to meet the Cowes sailmaker, George Ratsey. It was Ratsey who made the sails recommending and designing an innovative jibboom specifically for *America*.

Ratsey and Lapthorn rule the waves

As we have seen, for years the rights to make the sails for the British contender were shared between George Rogers Ratsey's Cowes loft and a sail loft established by James Lapthorn in Gosport in 1825. At first, Ratsey's firm was in the shadow of the Lapthorn loft, but such was his promise that the latter initiated a merger to form the long-lived firm of Ratsey & Lapthorn. It was no surprise, therefore, that in 1889 that the firms of Ratsey and Lapthorn merged to become the dominant force in making sails.

Thomas Ratsey (1851–1935) had a career that spanned the classic era of the America's Cup. His entry into the family business at the age of 15 heralded one of the most important contributions to America's Cup sailmaking made by a single individual. He was directly involved in seven challenges and the firm he controlled supplied sails for ten challengers and four defenders during his lifetime. Following the merger with Lapthorn, Tom Ratsey was then personally responsible for the sails of every challenger until *Shamrock IV* after his first involvement crewing on *Livonia* at age 20. His continuous involvement with the Cup began with the *Thistle* challenge of 1887, when his close friend G. L. Watson involved him in his designs at an early stage; his presence in New York during that challenge laid the foundations for Ratsey & Lapthorn's expansion in the United States. Allegedly, in 1833, when asked about the success of Ratsey & Lapthorn, Thomas was heard to respond, 'There is only one standard of work in this loft, that is the very best.'

Ratsey & Lapthorn's New York loft

Over the next four races from 1893–1901, every British challenger carried Ratsey & Lapthorn sails – *Valkyrie II*, *Valkyrie III*, *Shamrock I* and *Shamrock II*. All four races were won by the New York Yacht Squadron's entry. So, in 1902, recognising New York's significant competitive edge and an opportunity to dominate America's Cup sailing, Thomas Ratsey established a sail loft within Robert Jacob's City Island boat yard, New York, thus becoming the largest sailmaker in the world, employing over 180 sailmakers across its various sites. This enabled them to make sails for both the U.S. and UK America's Cup entries and developing a competitive yet amicable rivalry between their sail lofts.

Indeed, Thomas Ratsey's attendance at the 1895, 1899 and 1901 Cup races became more than the now expected attendance of the challenger's sailmaker. On all these occasions he took home significant orders from American yachtsmen who recognized his unique talent. By 1901 many of these were lobbying him to establish a loft in the U.S., which he did within Robert Jacob's City Island boatyard in

---

Above left: George Rogers Ratsey (1793-1850), who founded the Cowes loft.
Above right: Charles Ratsey (1812-1897).

1902. What resistance there was to the English invasion was effectively overcome with his firm's production of a near perfect mainsail for Cornelius Vanderbilt's New York 70, *Rainbow*.

This bold venture was given a significant boost when the same year the American railroad investor J.P. Morgan, became a supporter and early advocate of Ratsey and Lapthorn commissioning the New York Loft to make the sails for his private yacht *Corsair*. This is immediately followed by the decision of the New York Yacht Club to make the sails for the American America's Cup entrant and winner *Columbia* that year. A trend had been set with the Ratsey and Lapthorn New York loft building the sails for the next America's Cup defender, *Reliance*, with the Royal Yacht Squadron in the Isle of Wight commissioning Ratsey & Lapthorn's Hampshire sail lofts to equip its challenger, *Shamrock III*.

By 1917, George E. Ratsey (1876–1942) had decided to open a new loft on New York's City Island; it stretched over three floors, each over 150' long, making it the largest sail loft in the world. New strands of cotton from Sudan heralded in further innovation, driving demand in America. He was president of Ratsey & Lapthorn, Inc., City Island.

By 1934 the New York loft was making the sails for the United States America's Cup defender, *Rainbow*, whilst the Gosport loft equipped *Endeavor I*, the British entrant. Three years later, Ratsey & Lapthorn were again the choice on both sides of the pond, the New York loft making sails for the Cup defender and 'Super J' *Ranger*, whilst the UK loft equipped *Endeavor II*, the British entrant, with George Ratsey amongst the crew.

The Lipton Era boosts the reputation of Ratsey and Coker Canvas

In August 1898, with the United Kingdom Royal Yacht Squadron having suffered four spectacular defeats to the New York Yacht Club, the America's Cup organisers in the States were resigned to the prospect that there might not ever be another British challenger. Then, an unexpected telegram arrived from the Royal Ulster Yacht Club in Bangor, Northern Ireland, asking if they would accept a challenge from Sir Thomas Lipton. The challenge was accepted, ushering in one of the most illustrious periods in the history of the America's Cup and further enhancing the reputation of Ratsey & Lapthorn.

In the period 1899–1930 Lipton challenged and lost to the American holders of the Cup on five occasions under the aegis of the Royal Ulster Yacht Club. His yachts were all named *Shamrock (I–V)* and all carried Ratsey & Lapthorn sails. His well-publicised efforts to win the America's Cup earned him a specially designed cup for 'the best of all losers' and being rewarded with a posthumous induction into the Herreshoff Marine Museum/America's Cup Hall of Fame in 1993, to be joined by Thomas Ratsey's induction in 2009.

Lipton (1848–1931) was the second son of a poor Irish family from County Fermanagh that had emigrated to Scotland during the potato famine of 1845 that

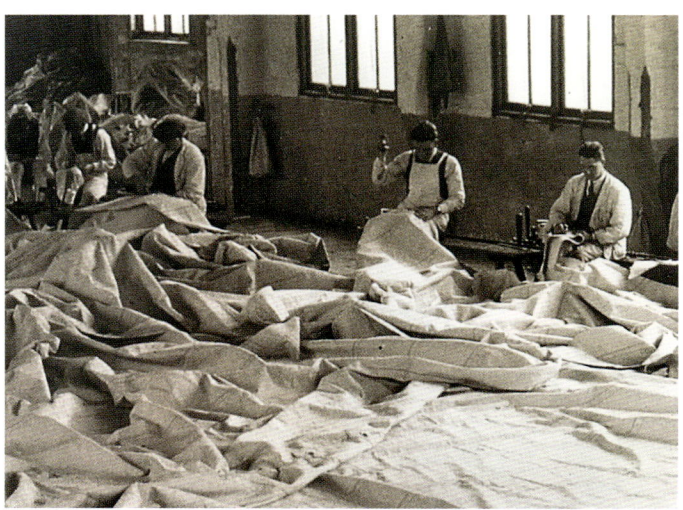

Above: Making sails in the Bermuda style rig for *Britannia* at Ratsey & Lapthorn's Cowes loft, 1931.

wreaked havoc across Ireland, settling in the Gorbals area of Glasgow. Leaving school early to help support his family, aged 14 years, he became a cabin boy on the Glasgow to Belfast package steamer, becoming captivated by life at sea and the tales of sailors. He soon headed off to work in the tobacco fields of Virginia (echoes of the early life of William Dampier) and rice fields of South Carolina before ending up as a clerk in a grocery store in New York. With $500 of savings, Lipton returned to Glasgow in 1870, initially to work in his family store before establishing his own shop.

Twenty grocery stores in Glasgow soon became 300 across Scotland. Lipton was clearly a successful fledgling businessman. By the mid-1880s Lipton had success with investments in projects in Nebraska and in 1888 opened a tea-tasting office. The freewheeling entrepreneur recognised the potential for growing the tea market in an age when tea was a rare and expensive luxury, he believed that anyone, of any class, should be able to enjoy tea at its best. Visiting British Ceylon (now Sri Lanka)

Above: *Vanity Fair* cartoon of Sir Thomas Lipton 'Shamrock', 1901.
Right: Part of cover of the U.S. satirical magazine *Puck*, from 14 November 1909, showing Uncle Sam goading Sir Thomas Lipton by holding the America's Cup and boasting about the number of times USA had won the 'Auld Jug'.

in 1890, he made various business deals including the purchase of Ceylon tea from local farmers.

He established the Thomas J Lipton Co.® tea packaging company in Hoboken, New Jersey and began to look for ways to make packaging and shipping less expensive. Instead of arriving in crates, loose tea was packed in multiple weight options. He also cut out the middleman and was the first to sell loose tea directly across Europe and the USA. Today, the Lipton Tea brand is owned by Ekaterra, Unilever's specialist tea division.

It was to be yacht racing that would be the vehicle to allow him to combine his love of sport with business, a celebrity lifestyle socialising with influential families on both sides of the Atlantic.

An inveterate sports fan rubbing shoulders with royalty

Lipton's interest in yachting was shared by King Edward VII and King George V. His interest in sport stretched beyond sailing, embracing both soccer and rowing. The Sir

Below: Preparing for the America's Cup, Barcelona 2024.

Thomas Lipton Trophy was a forerunner of modern-day European football competitions, created by Lipton with the Italian royal family. The competition was held only twice, taking place in Turin in 1909 and 1911. The trophy was awarded to the winners of the competition for invited teams from Italy, Germany and Switzerland. The Football Association refused to be part of the event, so Lipton invited West Auckland, an amateur team from County Durham, to take part. They won the inaugural event and successfully defended the trophy in 1911. He also initiated the Copa Lipton or Copa de Caridad Lipton, contested between the Argentina and Uruguay national teams. The competition was held 29 times between 1905 and 1992 by the two countries on either side of the Río de La Plata with the condition that the teams be made up of only native-born players.

In terms of rowing, the Sir Thomas Lipton Cup stands over three feet in height. Since 1914, it has been a symbol of success for oarsmen and oarswomen in the mid-western area of Canada and the United States, who are members of the North-western International Rowing Association and is awarded to the winning rowing club in the Association's annual championship regatta.

Above: Preparing for the America's Cup, Barcelona 2024.

Lipton's philanthropy was legendary. At Queen Victoria's diamond jubilee in 1897 he gave £20,000 to provide dinners for a large number of the London poor. During the First World War he supported the efforts of the Red Cross, the Scottish Women's Hospitals Committee and establishing hospitals in Serbia during the typhus epidemic of 1915.

War leads to diversification

Come 1942, as part of the war effort, Ratsey & Lapthorn diversified to equip the U.S. forces with custom canvas orders, including tank covers, plus, in conjunction with the long-established Connecticut Luders shipyard, they developed a rescue boat for downed airmen that was attached and deployed from the undercarriage of a plane.

Interestingly, this was a collaboration born out of a shared interest in yacht racing. Alfred (Bill) E. Luders Jr. was a racing sailor and Luders became known for its sailboats, the most popular of which was the one-design sloop, of which about 220 were built in the years following World War II using a new technique of plywood moulded with glues. Many of these sloops are still raced to this day. The most famous Luders sailing yacht was called *Weatherly*, which in 1962 successfully defended the America's Cup against the first challenger, *Gretel*, from Australia.

In 1958, Ratsey & Lapthorn's New York loft was still making the sails for the U.S. entrant in the America's Cup, *Columbia*, whilst the UK loft equipped *Sceptre*, the British entrant. Colin Ratsey sailed on board the American boat whilst George Ratsey crewed for the British in the race, which was won by *Sceptre*.

However, by the early 1980s, following the demise of the great sailing yachts and the introduction of modern synthetic fabrics, Ratsey & Lapthorn announced the closure of the New York loft but sailmaking continues in Cowes under the guidance and stewardship of its new owner and CEO, Jim Hartley. The focus remains the making of high-quality sails and the development of branded luxury canvas and leather travel luggage, with a sail loft in Barcelona.

As a footnote to this story, it is sad to announce that today sailcloth making in the United Kingdom is listed as 'endangered' on the Heritage Crafts Association 2021 list of traditional crafts at risk of extinction. However, the Coker Rope and Sail Charities are now collaborating closely with the Bristol Weaving Mill to revive the art and skill of making Coker Canvas using traditional methods. Ratsey & Lapthorn Limited continues to prosper and plans are emerging for a renewed relationship between the Cokers and the company.

Above: Preparing for the America's Cup, Barcelona 2024.

The Pirate and the Poet

Fragments of idolatry

In 2021, East Coker lost a favourite son and I lost a long-standing friend and neighbour. David Foot was born in 1929 at Verandah Cottage, East Coker, to Margaret and Frank Foot. He left school at 16 to become a junior reporter on the *Western Gazette* – Yeovil's local weekly newspaper – and was to become one of the best of cricket's legion of fine writers. Frank Keating, writing in *The Guardian* in 2009 to mark David's 80th birthday, wrote that, 'we were invited to raise a glass to the monarch of the counties. David loved his life in press boxes at cricket grounds of his beloved Somerset and at the football grounds of Yeovil Town and the two Bristol teams – Rovers and City. 'There will never be another David Foot', wrote Mayukh Ghosi, whilst *The Cricketer* opined, 'To read and re-read David Foot is to understand the height to which cricket writing might aspire.'

We often shared anecdotes about the characters who filled our lives in Coker. We reflected on the magical aspects of living in south Somerset. We discussed the pros and cons of a feudal community, of lords of the manors whose wealth was enlarged by their involvement in the slave trade, balanced with the fact that their beneficence was often shared with the community.

We wondered who the celebrity and mystery guests were often seen arriving at Coker Court, West Coker Manor or at North Coker House – rumours of their status fuelled local gossip.

David wrote of the 'illustrious visitors who enjoyed Coker Court's dignified grandeur'. They included two Liberal prime ministers – Asquith and Lloyd George – along with figures from the arts and literati, such as Harold Nicholson and Vita Sackville-West, who spent the first night of their honeymoon in the Georgian wing. Nearly 30 years later, the glamorous and wealthy Alice Keppel arrived. She had been the mistress of Bertie, Prince of

Above: View from Tellis Hill, East Coker.
Right: William Dampier.

Wales, later King Edward VII. Mrs Keppel's daughter, Violet Trefusis, followed her to East Coker. Violet, according to David, lived in a world of unconventional Bohemian standards, she inclined to sapphic behaviour and cross-dressing' and commenced an affair that was to rock the polite society of south Somerset and beyond... More of this later.

David often expressed deep concern that the story of Coker Canvas was untold and unheralded – for him it was deserving of greater attention. We also shared our mixed, confused views about the two historical and cultural giants of this story: William Dampier and Thomas Stearns Eliot.

In 2001, David published *Fragments of Idolatry: From Crusoe to Kid Berg – Twelve Character Studies.* It seems appropriate, therefore, to explore fragments of the lives of Dampier and Eliot, two globally important figures, from the perspective of those aspects of their lives that interconnect and are entwined in the story of Coker Canvas and who for two boys growing up in East Coker village were in the words of Davide Foot, 'the source of distant idolatry'.

The Pirate

Dampier, the pirate with the exquisite mind

William Dampier, known as 'the pirate with the exquisite mind', dined with Samuel Pepys, Sir Hans Sloane, John Evelyn and John Lock. Charles Darwin took Dampier's books with him on board HMS *Beagle*, and they also influenced Jonathan Swift's *Gulliver's Travels*. An oft-forgotten great explorer and innovator who was at sea some 80 years before Captain Cook, Dampier has been in danger of being relegated to a footnote in history, sandwiched between Sir Francis Drake's defeat of the Spanish Armada in 1588 and eclipsed by Cook's celebrated voyages. Yet, without Dampier, *Robinson Crusoe* would not have been written.

Arthur Mee's description of East Coker in *The King's England: Somerset* (1968) opens with a sentence that rather stretches the supposed links with the Daniel Defoe story (Defoe would actually have been familiar with this area of Somerset, having fought in the Battle of Sedgemoor during the English Civil War) when he writes, 'It was an East Coker boy who piloted the ship which brought Robinson Crusoe home.' Having peeked the reader's interest, Mee takes us to the brass plaque on the north wall of the village church on which is inscribed:

Buccaneer, Explorer, Hydrographer
Thrice he circumnavigated the Globe, and first of all Englishmen, explored and described the coast of Australia.

An exact observer of all things in Earth, Sea, and Air, he recorded the knowledge won by years of danger and

hardship in Books of voyages and a Discourse on Winds, Tides and Currents which Nelson bade his midshipmen to study and Humboldt praised for scientific work.

Born at East Coker in 1651

He died in London 1715

And lies buried in an unknown grave

The world is apt to judge of everything by the success, and whosoever

has ill fortune will hardly be allowed a good name.

Above: Bridge Farm (now Hymerford House), William Dampier's birthplace.

The most extraordinary of lives summarised in just a few lines.

The enlightened members of the Royal Society respected him. He was in the foment of ideas and information that found fertile ground in the coteries and cliques of the Restoration. The *Australian National Dictionary of Biography* calls him 'a curious man for a curious age, who lived a swashbuckling life with pioneering scientific achievements.'

Writing in *The Guardian* newspaper in August 2004, Kevin Rusby provided this summation of the life of the man: 'Dampier set out as a buccaneer but ended up a friend of the brightest luminaries of Restoration London. Who was the first Englishman to eat mango chutney? Who was the first to describe chopsticks and help hundreds of other words to the English language? It was the same man who circumnavigated the globe an unprecedented three times, who first correctly described the winds and currents of the Pacific and who became the first Englishman to land on the Australian mainland.'

William Dampier was the second son of a tenant farmer, George Dampier, and his wife, Anne, who both died when Dampier was a young boy. He was born in East Coker at Grove Farm (originally assigned to Richard de Argentine, a close friend of William the Conqueror). Today, this proud, thatched, Grade I listed property is now more commonly known as Hymerford House. It was originally the main managerial farmhouse for the Helyar estate being centrally located, close to the mills on the Coker Water that were producing flour as well as those producing flax and hemp.

Dampier was baptised in St. Michael and All Angels' church. Orphaned in 1658 with the death of his father, who left him £100 worth of farmland, his guardians included his great uncle, his grandmother's brother, the sailcloth maker John Giles. In 1634 Giles, also a tenant of the Helyar estate, had produced the sails for *Sovereign of the Seas*. So, at seven years old, the young Dampier was already infused with aspects of sailcloth making and the lure of the sea.

He was formally educated under aegis of Squire Helyar, who sponsored his education at Yeovil and then Crewkerne Grammar Schools (Crewkerne being the alma mater of Admiral Thomas Hardy, captain of HMS *Victory* at the Battle of Trafalgar) before enrolling at him King's School in Bruton. The Helyar family library at Coker Court would have fuelled Dampier's interest in the natural world. It included the 14th-century *Liber de Hirbis*, one of Manfredus de Monte Imperiali's illustrated Latin catalogues of plants – books that I could gaze upon, but was not allowed to touch, during my many childhood visits to Coker Court.

The incorrigible romancer

Even at the age of 14 he appears to have been an 'incorrigible romancer, a rover by disposition.' It was his one wish that he went to sea and started his seafaring life when Archdeacon Helyar arranged for him to be indentured to a ship's captain in Weymouth, setting off on a 4,000 nautical-mile (7,408km) voyage to Newfoundland and Labrador. He did not enjoy the freezing waters and cold weather, so he headed back to London. Upon his return, he found passage in 1670 on the East Indiaman, *John and Martha*, bound for Bantam in Java. On this voyage, Dampier gained experience in navigation, particularly the calculation of longitude and observation of wind and weather patterns.

The *John and Martha* returned to London, then, after just two months in *Bantam*, Dampier enlisted in the Royal Navy in 1672. A year later he set off on what was to become the most adventurous and eventful of lives. He was posted to HMS *Royal Prince* and was to take part in fierce maritime encounters during the Third Anglo-Dutch War (1672–1674). In August 1673, he was present at the Battle of Texel that involved 207 warships and incurred a loss of over 3,000 men and boys. It was no wonder that he quit the Navy.

In 1674, after recovering from illness, Dampier accepted an offer to manage the Bybrook sugar plantation in Jamaica, owned by William and Cary Helyar. It was a fateful decision because it directly led to Dampier's buccaneering life on the high seas. Dampier and his employer, William Whaley, the manager of Bybrook, clashed bitterly over the management of the plantation. As a result, Dampier soon left Bybrook to try his luck in the potentially lucrative logging industry on the shores of the Bay of Campeche, at the southern end of the Gulf of Mexico.

The purple-red dye from logwood trees commanded a high price in London, but Dampier soon realised that any fortune to be made would not come easily since England and Spain were in conflict over control of the logging industry. His time as a logger, however, contributed to his

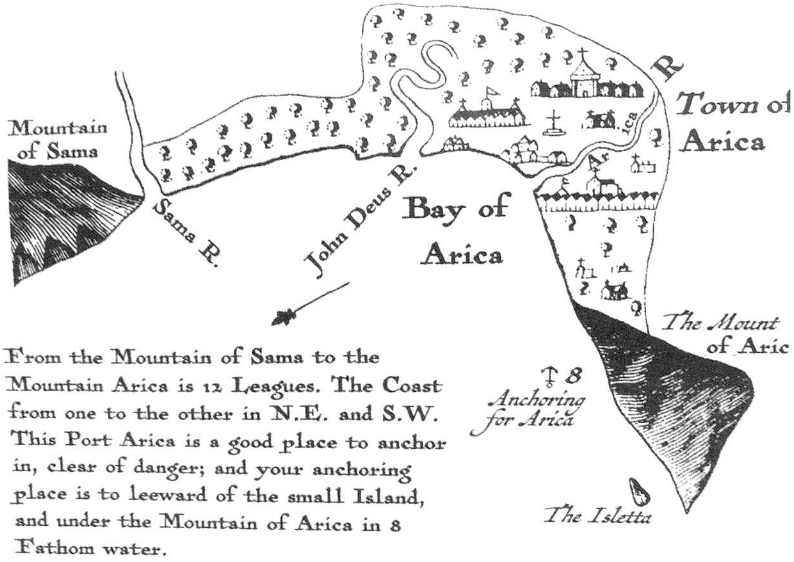

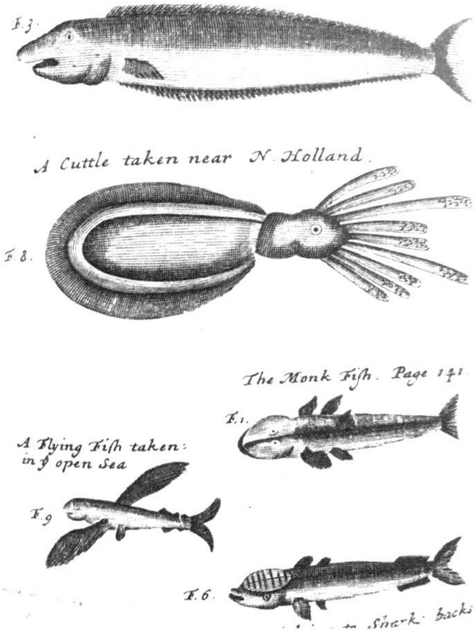

'This Birds Eye is of a Bright Red'

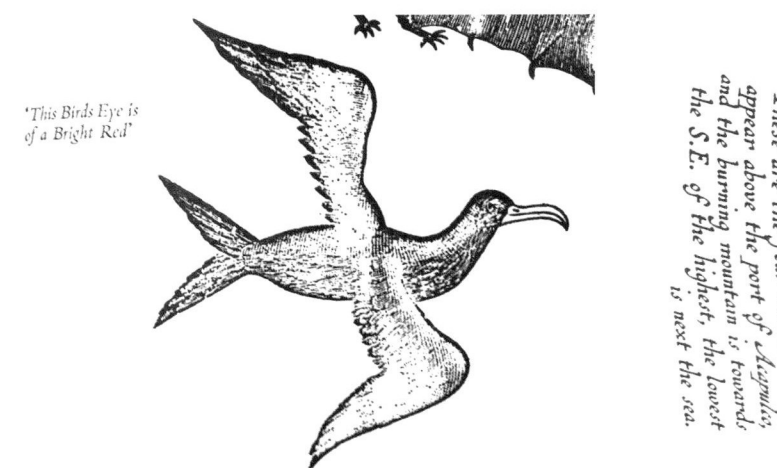

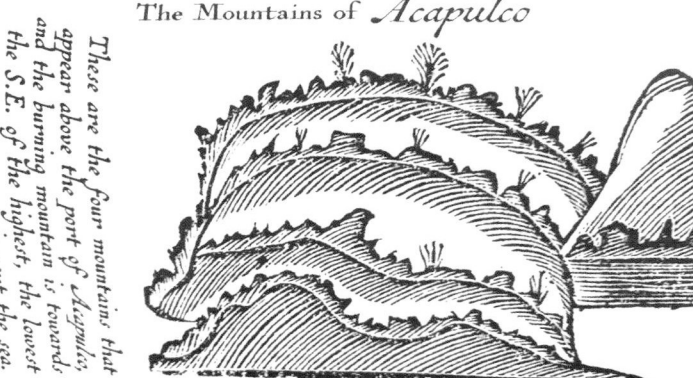

Above: Various drawings by Dampier from *Dampier's Voyages*, edited by John Masefield, 1906.

impressive ability to describe nature. He kept meticulous journals throughout his life, starting with his ethnographic observations in the Bay of Campeche of monkeys, spiders, snakes and the armadillo. Dampier's drawings and the introduction of the term 'subspecies' into the English language later influenced Darwin's theory of evolution.

Once again, his stay was short lived, and within 12 months Dampier was on his way, leaving behind his logging life after a hurricane hit the Bay of Campeche in June 1676. However, his detailed account of this hurricane has been recognised by scientists as the 'first accurate description of this phenomenon'.

A Caribbean drifter

Having no money, like many other Caribbean drifters, Dampier slipped into a life of freebooting and buccaneering – plundering became a source of income. In the period 1680–1720 the business of piracy was not as we imagine. Dampier and his fellow privateers were amateurish, eclectic in their interests and mostly inoffensive. When Dampier's travel journals were later published to critical acclaim, he would, as one biographer explains, 'downplay his buccaneering career, preferring to reframe his account as one of scientific enquiry, which through unfortunate circumstances, had to be conducted in the company of drunken thieves and ruffians.' Yet, the English establishment viewed buccaneering as a low-cost way of countering the Spanish. Buccaneers were often accepted into high-society, for example, Sir Henry Morgan (c. 1635–1688), the ruthless Welsh buccaneer, was elevated to knighthood in 1674 by King Charles II of England (1660–1685) and appointed Lieutenant-Governor of Jamaica, bringing him directly into contact with the Helyar family.

Dampier the buccaneer was different. In his meticulous journals he recorded sights, sounds, tastes and peoples along with careful observations of winds and currents. It was said he was no gentleman jotter, often penniless, often alone, but always willing to move on. In 1683 he seized a Danish slaver ship off the coast of Sierra Leone and renamed her *Bachelor's Delight*! Two years

later Dampier jumped ship to join Captain Swan in the *Signet* (nicknamed the *Cygnet*, a pun on Captain's name), exploring the southern oceans. In 1688 they sighted the north-western coast of Australia and sailed into King Sound (near present-day Broome, Western Australia), where Dampier ungenerously recorded that the indigenous people 'were the miserablist people in the world', before sailing onwards to Sumatra and Christmas Island.

Above: Prince Jeoly.

On the way home in 1691, William Dampier shamefully became a slave trader. He purchased a heavily tattooed man at Fort Saint George (India), known as the painted Prince Jeoly (sometimes Giolo), who was originally from Miangas (sometimes Moangis, present-day Palmas, North Sulawesi). Ever the journalist, Dampier described Prince Jeoly as being 'painted all down his breast, between his shoulders behind; on his thighs (mostly) before; and in the form of several broad rings or bracelets round his arms and legs' that he disgracefully sold for 'novel exhibition'.

He did not care a fart

The next six years were spent writing *A New Voyage Round the World,* which was published to acclaim in 1697. It was based on his many journals, and he also became reacquainted with his wife, Judith, whom he had not seen for 12 years. His book was an immediate bestseller and is said to have influenced both the English poet and literary critic Samuel Taylor Coleridge's (1772–1834) poem *The Rime of the Ancient Mariner* and Jonathan Swift, whose fictional character in *Gulliver's Travels*, indeed, Captain Lemuel Gulliver, was said to be based on Dampier's experiences.

A New Voyage Round the World – later to be re-edited by the Poet Laureate John Masefield in 1906 – brought Dampier to the attention of the British Admiralty and the Royal Society, who invited him to give lectures on his adventures and his visit to New Holland. As a result, the Government requested he command the first scientific research expedition to New Holland.

Consequently, in January 1699 he was appointed captain of the 300-ton HMS *Roebuck* with a crew of 50 men to explore New Holland (Australia); by July, the *Roebuck* had landed at Shark Bay in north-west Australia. However, Dampier's lack of knowledge of Naval protocols soon brought mutinous intent amongst the crew. 'Almost weary', he headed for Timor then on to New Guinea (then Batavia) for repairs before setting sail for England via Cape Town and St Helena (February 1701) and Ascension Island where the ship was found to have sprung a leak and sank. The abandoned crew and their captain were rescued by a passing ship. Once back in England in 1702 he was found not to be fit to command a Royal Navy vessel following accusations of incompetence made by one of his former lieutenants. Interestingly, Dampier had noted in his journal that this deputy, Fisher, 'did not care a fart for a non-Navy captain and was ready to make trouble.' Trouble indeed, as Dampier was court-marshalled and considered 'not a fit person to be employed by the Queen.'

This setback was soon forgotten when, in 1702, Dampier was given command of the ship *St. George* by its private owner. Along with the command came letters of marque allowing him to capture enemy ships for the benefit of the owner in the South Seas – a fighting journey, not one of discovery. One of his lieutenants, Alexander Selkirk, was thrown off the *Cinque Ports* – a ship accompanying the *St. George* – and was marooned on Juan Fernandez Island, an episode that formed the basis of Daniel Defoe's 1719 novel *Robinson Crusoe*.

Dampier published extensively between 1700 and 1709. His supplement to *New Voyage*, titled *Discourse of Trade Winds*, was recognised immediately as a 'most valuable of pre-scientific essays', together with a two-part account, the *Voyage to New Holland*. Despite this acclaim, on his return to England in 1707 after five years away, his reputation as a sea captain was in tatters. Undeterred, and surprisingly, the following year Dampier became a pilot on a privateer vessel, *The Duke* (ironically the ship that eventually rescued Selkirk in 1709), which rounded Cape Horn and raided Spanish towns on the west coast of South America.

This final expedition was a rich one. A Manila galleon, the *Nuestra Señora de la Encarnación y Desengaño* – which translates to 'Our Lady of the Incarnation and Disappointment' – was captured off the west coast of Mexico by Dampier and the privateer fleet on New Year's Day, 1710. Its cargo consisted of musk, cinnamon, Chinese porcelain, silver plate and jewels, a prize bounty estimated at over £200,000, of which just £1,300 was allocated to Dampier.

Deserving of greater recognition

Dampier's life sank quietly in the end and into debts of £2,000. Without honours, without even the actual date of his death being recorded, he passed away on his cousin's farm in Dorset, near to East Coker, where his illustrious journey had begun. His work is celebrated in Australia and New Zealand; indeed, across the southern hemisphere many places bear his name. Dampier the naturalist and hydrographer deserves greater recognition. He was treated badly by parts of the English establishment because, for many, he was too wayward and individualistic to become a hero in the style of Captain Cook.

Dampier should be remembered as Australia's first naturalist, the first European to undertake a scientific study of Australian flora and fauna. His published travel accounts also introduced into the English language over 1,000 new words, including albatross, banana, free trade, intelligence, avocado, flamingo, cashew, chopstick, catamaran, barbecue and breadfruit. Dampier was the finest seafarer and navigator of his time. Despite his buccaneering years at sea, he had little to show in the way of profits, and he only had his journals and collected specimens.

Indeed, Dampier's more recent editor, Gerald Norris, says of the man, 'there is growing recognition of his accomplishments. His three books rank as more than superlative travel literature: they also have scientific importance. His object was to see all countries and observe the works of Nature'. He is acknowledged as a pioneer of scientific exploration, commemorated in the Dampier Archipelago (Indian Ocean), Dampier Strait (Papua New Guinea), Port of Dampier (Western Australia) and Mount Dampier (New Zealand).

A distant idol, flawed but inspiring

Dampier's life was one of conflicting achievements. It is easy to understand why David Foot and I regarded him as a distant idol. He was a brilliant, self-educated scientist, but he was severely flawed. Many of his exploits caused his backers to suffer financial loss and reputational damage and his leadership skills left much to be desired. His nautical contemporaries remarked about his moods and misjudgements that led to insubordination by his crews at sea. 'His eyes,' as David Foot once said to me when we were discussing Dampier's portrait in the parish church, 'were dark and hard. Sadly, they belonged to a plunderer as much as to the observer of natural history. They were the callous eyes of a man whose savagery, racism and occasional sadism cannot be denied. There is the undeniable paradox of a genius who respected the natural world yet was to order the public flogging of a Black woman taken on as a cook who was too wanton with her bodily favours and who unashamedly sold Prince Jeoly for public show.'

Masefield, after editing *Voyages* in 1906, did much to promote the extraordinary contribution that Dampier had made to our understanding of the natural and scientific world. It was these texts that helped inspire this Poet Laureate and erstwhile sailor to write his 1910 collection of salt-water poems and ballads. Masefield wrote extensively about the sea and sea adventures as being a strong influence on the English people and to humanity as a whole. His most famous of the sea-theme poems is *Sea Fever*, with the oft quoted opening lines: 'I must go down to the seas again, to the lonely sea and the sky, And all I ask is a tall ship and a star to steer her by...'

Masefield also pays homage to hemp and flax in his poem *The Ship and her Makers*, recognising the significance of these plants in the whole story of sailing he devotes an entire section of the poem to the hemp and the flax:

We were a million grasses on the hill,
A million herbs which bowed as the wind blew,
Trembling in every fibre, never still.
Out of the summer earth sweet life we drew.
Little, blue-flowered grasses up the glen,
Glad of the sun, what did we know of men?

Above: T. S. Eliot.

The Poet

T. S. Eliot, the Nobel Prize winner and Britain's favourite poet

T. S. Eliot, poet, essayist, publisher, playwright, literary critic and editor, is considered one of the 20th century's major poets and a central figure in English-language Modernist poetry. Born in St. Louis (Missouri) in 1888, he moved to England in 1914 at the age of 25 and went on to settle, work and marry here, becoming a British citizen in 1927, subsequently renouncing his American citizenship. Voted Britain's favourite poet in a BBC poll in 2009, he is memorialised in both Poets' Corner in Westminster Abbey and in the Parish Church in East Coker. Eliot first attracted widespread attention for his poem *The Love Song of J. Alfred Prufrock* in 1915. It was followed by *The Waste Land* (1922), *The Hollow Men* (1925), *Ash Wednesday* (1930) and *Four Quartets* (1943). He was awarded the 1948 Nobel Prize in Literature, 'for his outstanding, pioneer contribution to present-day poetry' and is regarded as the father of modernist poetry.

The return journey, the closing of a circle

Eliot first came to England in 1914 as a postgraduate student at Merton College, Oxford. Settling in London, he was to publish his most significant poem, *The Waste Land*, in 1922. Five years later, this avid churchgoer was baptised and confirmed in the Church of England. According to Paul Keer, Chair of the T. S. Eliot Society, Eliot described himself as 'a classicist, a loyalist in politics and an Anglo-Catholic in religion.' It was as church warden at St. Stephen's Church on the Gloucester Road that he embarked on the *Four Quartets*, which were written over the period 1935–1942.

Although, as we will see, the poem *East Coker* is seen as a metaphor of an idyll of a lost life, Keer believes that 'it is likely that Eliot's decision was about the closing of a circle. He saw his family coming from British stock.' Indeed, the Eliot name first appears in the church register in 1563, three years after the extant register. The last record of the family name occurs in 1664 recording the marriage of a Mary Eliot.

The poem pursues the theme of time past and time present. As Paul Keer points out, *East Coker* begins 'In my beginning is my end', which is an inversion of the motto of Mary Stuart (Mary, Queen of Scots), taken from the work of the 14th-century French poet Guillaume de Machaut, *'En ma fin est mon commencement'* – 'my end is my beginning' – which ends *East Coker*. The circle is complete.

Eliot's first known visit to his ancestral home of East Coker in 1937 was followed two years later by what would be his last before his death in 1965. On the first visit he travelled to Yeovil from London by train then took a taxi to East Coker. He was enthused with the images and landscape from reading Thomas Hardy. The 1936 visit led to a letter that described East Coker as 'delightful' and 'pretty' and that Eliot felt 'at home' there. He also wrote that the village had 'a sort of Germelshausen effect', referencing a German

tale of a traveller who comes across an ancient village with church, gravestones, farmers and dancers, itself a folktale that Keer says is 'a story of time, and lost love which cannot be regained; a story that may have influenced Eliot's imagery in "East Coker"'.

On the first visit he stayed as a guest of the Heneage family at Coker Court, whilst on the occasion of the last visit he stayed in West Coker Manor as a guest of Sir Matthew Nathan. During these visits there must have been great discussions about the history and lore of the Cokers. It was about this time that Violet Trefusis, one of the Bloomsbury Group writers, a friend of Virginia Woolf, lived in one of the wings of Coker Court and then moved to stay with Nathan in West Coker Manor. Trefusis is chiefly remembered for her lengthy affair with the writer Vita Sackville-West that both women continued after their respective marriages dissolved. The scandal of the affair was featured in novels by both parties and in Virginia Woolf's novel *Orlando*, and she may have been the inspiration for aspects of the character Lady Montdore in Nancy Mitford's *Love in a Cold Climate*. Eliot, Trefusis and Nathan – what interesting dinner table conversations.

Above: Bath's Bus, 'timeless transport'.
Right: Margaret and Frank Foot, Verandah Cottage, East Coker, c. 1985.

On this last visit in 1939 he was to meet the charming, chatty and patient Estate gardener, Frank Foot (my 'Uncle Frank'), who, as the church sexton, climbed the church tower twice a day to wind the mechanism to run the church clock every day for over 60 years. On the day in question, he was at his usual task in the church. According to David, Frank's son, 'My Father was forever vague about his conversation with the stranger. He said he [Eliot] seemed a nice sort of bloke. Bit religious, too – you could tell that. We had a word about football and the Yeovil & Petters team [later to become the famous Yeovil Town Football Club].'

Eliot was certainly regarded as a curiosity as he wandered the lanes and fields of East and West Coker – neat, formal, suit and tie, donnish in appearance. After this conversation, Eliot walked down the hill from the parish church, past the Helyar Tudor almshouses, to catch Mr. Bath's 'ever rattling green and white Bedford bus back to The Borough in Yeovil. As David Foot was to observe, 'timeless transport fitted his pattern and needs'.

A sense of wonder

It was clear that Eliot barely knew East Coker, having to familiarise himself with the place and the people,

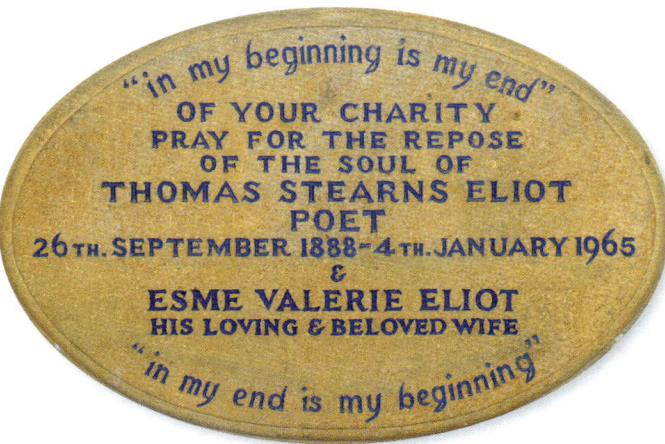

Eliot's family is recognised in *Burke's Distinguished Families of America: The Lineages of 1600 Families of British Origin Now Resident in the United States of America* (first published in 1826), which states that Andrew Eliot, a Calvinist of East Coker, Somerset, first settled in Salem, Massachusetts, then in Beverley, where he became the town clerk in 1690. Andrew, the so-called first 'American' Eliot, played a very prominent and infamous role within the colonial administration of the state. At that time, Massachusetts was an unceded territory of the indigenous Naumkeag people of the Massachusetts tribe. He was both witness to the dispossession of the Naumkeag lands ('The Indian Deed of the Town of Sale, 1686') through legal and financial coercion, and *Savage's Genealogical Dictionary of America* (1826) identifies Eliot as one of the jurors in the Salem Witch

but these visits clearly had an immense impact on him, providing the raw material around which he could weave his abstract thoughts and ideas. It was a revelation and epiphany for him after he discovered that his Elyott ancestors went from here to America (a Henry Elyott was a tenant of the Helyar family).

East Coker had been hit hard by the 1645 Plague and suffered further in the English Civil War. The village and its Manor were staunch Royalists and suffered the severe consequences and retribution, with Cromwellian destruction of much of value in the parish church and cemetery. After the end of the English Civil War, the Puritan Rebellion and the reinstatement of the Monarchy in the 1660s, Eliot's forebears, led by Andrew Eliot, a cordwainer (leather worker or shoemaker) from East Coker ventured to the New World, joining the colony established by the Puritan lawyer John Winthrop, who by the time Andrew Eliot arrived was Governor of the State of Massachusetts. T. S. Eliot's grandfather, William Greenleaf Eliot, was to relocate the family to St. Louis, where Eliot was born in September 1888.

Above: T. S. Eliot's Memorial at St. Michael's Church, East Coker.
Right: East Coker boundary sign on Ham stone.

Above: East Coker Church of St Michael and All Angels'.

Trials of 1692, stating he was 'of the juries, say tradit. wh. Tried the witches (of Salem) and had great mental affliction on that acc.in the residue of life.' Andrew Eliot was to describe the decision as a 'heinous crime'.

Eliot closely identified with another ancestor, Sir Thomas Elyot (1496–1546), the grandfather of Andrew Eliot, a highly respected Tudor scholar, writer and diplomat who produced the first Latin to English dictionaries. Eliot draws upon Sir Thomas Elyot's 1531 book on how to become a statesman, *The Boke Named Governour,* when describing the 'daunsinge and matrimonie' of the villagers. In this way, according to the author of the *East Coker Neighbourhood Plan*, Eliot 'poetically incorporates the familial inspiration from the past within the tradition and the bloodline of which Eliot felt part of and wrote.' Eliot clearly felt a personal relationship with Sir Thomas and

Above: Stained glass window, East Coker Church.
In his will, Walter Graeme Eliot of New York bequeathed $1,000 to install this stained glass window in East Coker church in 1936, commemorating Andrew Eliot and his descendants. T. S. Eliot disliked it intensely. In 1936 he wrote of East Coker that 'The only ugliness in the place is a stained-glass window in the church, recently erected by some cousin of mine – the most hideous I have seen.'

placed his vision of the village of East Coker as an actual place in the early Tudor period.

An English rural idyll

In 1940, *East Coker* was published. It was to be the first of the *Four Quartets* to have been written but appears as the second in the final published sequence. Originally featured in the Easter edition of *New English Weekly*, it was so well received that it was republished in the May and June editions that same year before being released as a pamphlet by Faber & Faber in September, selling 12,000 copies.

Four Quartets incorporates the four seasons and the four elements. Appearing as the second of the four, *East Coker* represents summer and the earth. For Paul Keer, the poem profound and complex, it is a 'mediation on time and timelessness, on history and salvation, on life, love and language woven into a musical structure.'

The poem draws upon his perception of an English rural idyll – he observed that, 'the village retains much of its secluded air ... the still point in a turning world ... time present and time past are both present in time future' in this traditional ordered society in danger of being lost. Some consider the *Four Quartets* to be war poems and part of Eliot's war effort. This has particular significance as *East Coker* was written just before America became involved in the war. The poem also contains references that reflect Eliot's respect for and acknowledgement of the importance of soil: 'earth feet, loam feet.'

A year later Eliot wrote *The Dry Salvages,* which would become the third in the set of *Four Quartets*. A poem about the sea and landforms ends with the line 'the significant soil', re-affirming his persistent and underlying interest in geology and soil as the fundamental feature of the human and natural lifecycle that is a common thread throughout the *Four Quartets* – ashes to ashes, dust to dust, and 'the parched eviscerate soil.' Indeed, he has been often referred to as 'the poet of the soil'. Eliot was drawn to the idea of

organic land management and husbandry in the 1940s echoing and influenced by his observations of farming in the Cokers.

According to Simon Jenkins, writing in *The Guardian* in April 2007, 'the village was to Eliot an idea, a metaphor to put to poetic use, an idyll of England at the start of the second world war. To an expatriate it was also soil, roots, something to which, however much he ignored it, he should dutifully return. ... 'in my beginning is my end, in my end is my beginning.'

However, Jenkins is critical of Eliot's representation of the village, noting 'Eliot's "frigid stare",the absence of happiness, the narcissistic gloom. He [Eliot] is like some Somerset Puritan hurling down hellfire on the wretches who have just staggered in from the fields.' All of this is unfair to this fair community. As Jenkins concludes, 'This is no place for Eliot's grim poem. East Coker should wear a perpetual smile on its face – and it does.'

For Eliot, however, he wrote after *East Coker* appeared that 'I suppose it will seem to some people a sombre poem, but to me it was a hopeful one, for the individual spirit not for society'.

Eliot remembered

Today, Eliot is remembered in East Coker by a humble oval plaque of Ham stone, simply engraved in a colour described by the author Rosy Cole as being 'as blue as the quivering Somerset flax.' The plaque sits above his ashes, which were interred on Easter Sunday, 1965 in the north-east corner of the golden-toned hillside of the parish church of St. Michael and All Angels'. It was always about symmetry for Eliot, and it was his choice to have his ashes interred in East Coker. The plaque is engraved with the simple phrase, 'Of your charity pray for the repose of Thomas Stearns Eliot' and the closing sentences of *East Coker*: In my beginning is my end ... In my end is my beginning.'

The resting place is in the northwest corner of the church, sitting underneath a stained-glass window donated by the Eliot family and recently installed in 1936 before his first visit by, as Eliot remarked, 'some cousin of mine', to commemorate Andrew Eliot and other descendants. Eliot's decision on this spot as his

Top left: The *Four Quartets*.
Above: Lectern, East Coker Church.

The Pirate and the Poet

final resting place seems not to have been influenced by the presence of this family memorial, even though he described the window as a 'blemish', noting that even the coat of arms was wrongly inscribed.

As a result, some suggest that Eliot's choice was to show respect for and be near the 13th-century effigy of a lady of the de Mandeville family (creating a strong link to the heritage of the village – what Eliot referred to as 'generational time') and the memorial (erected in 1908) to William Dampier. Eliot would have known that another Poet Laureate, John Masefield, had edited and published a two-volume edition of Dampier's *Voyages*, helping Dampier's rehabilitation. Indeed, two of the many words that Dampier introduced to the English language, 'porpoise' and 'petrel', were used in *East Coker*, aping the use of 'gull' and 'whale' in Masefield's 1902 poem *Sea Fever*, and in this way both Eliot and Masefield honoured the influence of Dampier on the English language.

Eliot's work was a strong influence on the Yorkshire-born poet, playwright and popular children's author Ted Hughes – later to become Poet Laureate in 1984. On Eliot's death in 1965, Hughes wrote to his editor at Faber & Faber, Charles Monteith stating that 'It was a great blow, … It's left things feeling extremely windswept. I for one lived very much under his eye, inwardly.' At the unveiling of a Blue Plaque to Eliot in London's Kensington in 1986, Hughes commenced his speech by saying, 'We are here to pay a small and simple tribute to a great poet. Eliot moves at large throughout all variations of language and culture, claimed by all, as they become aware of him, and needed by all'. Later, acknowledging that the poems in his 1957 collection *The Hawk and The Rain* take direct inspiration from *East Coker*.

In autumn 2023, as part of the Od Arts Festival, the organisers, OSR Projects of West Coker, commissioned Jack Young to write East Coker 0.2 after T. S. Eliot's second poem from *Four Quartets*. Young's response is exciting and challenging. It invites a new and contemporary perspective of Eliot and the original poem, opening with:

Turn your back on the scene tom
look upon it through tainted glass
see what you want to see
(try) and feel what you want to feel
say what you want to say
though words are always slipping

Young pulls no punches, ending East Coker 0.2 with:

From sea to land, land to sea
Pull the wool over the Helyar's eyes owlers cloaks
Elsewhere (within law) wool transported to the backs of
Helyar's people – enslaved
Bybrook, Jamaica, Helyar's crops
Plantation cruel.

Above: The New Inn, East Coker in 1908, later to become The Helyar Arms in 1948.

Musings and Mysteries

More questions than answers

There are a number of important questions still to be answered about the global phenomenon that occurred in the villages of East, North and West Coker. Why and how did these inland villages develop such a strong maritime focus for their economic survival? Were there factors beyond having the right natural resources (soils and water) and a well-developed spirit of enterprise?

Was it about local landowners and businessmen and women who had strong personal connections with national decision makers who were able to exert influence on governments, trade and investment? How did they punch above their weight consistently for so many years? What was it about the society, culture and environment of this place that nurtured such enterprise, creative talent and desire to succeed?

about commerce and wealth or was this support for local enterprise and business stimulated by a kind of inherent loyalty to the Cokers, their place of birth, residence or education?

What is clear is that, for some reason, the area certainly had links with more than its fair share of England's great seafarers and mariners. In addition, there is the enduring mystery of the origins of the 'maritime fields' and why the Cokers belonged in the spirit of maritime adventure for centuries. Threading through the loom that weaves together the different stories in this history of the villages are the most extraordinary number of coincidences where stories collide, where the impacts of a decision ripple through the centuries, where serendipity creates moments of despair and great opportunity. It is an enduring story that is sustained through the America's Cup to this day.

The maritime roll of honour

These maritime connections are impressive. As well as William Dampier, it includes Thomas Hardy, captain of HMS *Victory* at the Battle of Trafalgar, who was a pupil at nearby Crewkerne Grammar School; Nelson's brother, the Reverend William Lord Nelson, and Sir Amyas Preston, who came from nearby Cricket St. Thomas; Admiral Alexander Hood, who hailed from Bridport (1776–1814); Vice Admiral Sir Samuel Hood (1762–1814) of Compton Dundon; Sir Walter Raleigh, a regular visitor who lived a few miles from East Coker; John Hanning Speke (1827–1864), explorer, was from Ilminster; General at Sea Robert Blake (1598–1657) came from near Bridgwater; Captain John Marchant (1540–1592), who sailed under Sir Francis Drake, was born in Yeovil, and in East Coker church cemetery there are graves to Henry Michell of the Royal Navy (1830–1853) and Commander John Bullock Michell R.N. (1829–1868).

Was the success of this maritime heritage driven by natural leaders in the community or was it something much more organic in its development? There was certainly evidence of significant levels of local and national patronage, but why did so many influential landowners, politicians and the upper echelons of the Royal Navy become so involved in the sailcloth and twine making industries of East, North and West Coker? Was it purely

Left: Coker Court and the Parish Church, East Coker.

Above: S. Guppy working the webbing looms at Drake's factory, c. 1965.

The area's maritime procurement history dates from the 13th and 14th centuries, when the Coker manors were held by the de Courtenay Earls of Devon, who were close blood relatives of the Plantagenet Lancastrians and the Tudors – especially Edward III, Richard II and Henry IV, V and VI. The de Courtenays were one of the most important English Renaissance families and significant political players on the national stage throughout the 13th–15th centuries. The Lords of the Manors were interested in exploration of the world, they were active in colonial projects and supported privateers.

During this time, Naish Priory (now a Grade I listed building) in East Coker was built and, in 1397, Sir Phillip Courtenay came to live there. At the time he was Richard II's Lord Lieutenant of Ireland and High Admiral of the Western Seas (Emery, 2006), making the family the supreme naval commanders with responsibility for the defence of the south coast of England and transport logistics during the Hundred Years' War. The de Courtenay's long connection with the Cokers is referenced by T. S. Eliot in *East Coker* as 'the house of succession'. In 1941, Arthur Mee also referenced the importance of the maritime links, in his book on Somerset in the series The King's England, calling this area 'Dampier's village'.

As a supplementary question, why did this relationship with the sea, especially with the Royal Navy and the Merchant Navy, persist over so many years well into the 20th century? On a personal level it is interesting to note that over three generations before 1950 every male member of the Stevens family, and indeed my mother and her family, joined either the 'Senior Service' or the 'Wavy Navy'.

Where do good ideas come from?

The second question has a more generic focus and has exercised minds for a long time: where do good ideas come from, and how do they appear and gestate, especially in a remote rural area with a small population?

The research of Steven Johnson, summarised in his work entitled *The Natural History of Innovation,* sheds light on this conundrum, as he identifies different conditions, or key environments, which can be the catalyst for the nurturing of good ideas, concluding that, 'we are better served by connecting ideas than by protecting them.' This certainly seemed to have been the modus operandi amongst entrepreneurs operating within the Cokers, with the examples he identifies in his monograph indicating that at least six of Johnson's principles for innovation were undoubtedly 'at work', ensuring the successful growth and evolution of the Coker rope and sail industry. These six principles are:

1. The importance of 'liquid' networks – engaging with people from different backgrounds, skill sets and businesses to broaden the scope and range of thinking needed to innovate and find new solutions.

2. The 'adjacent possible' and the willingness to talk to the person closest to you within your community – they may well have the answers derived from other experiences they have had.

3. The need to be more counterintuitive and notional, allowing intuition and hunch to guide the way forward in balance with rationale and evidence.

4. The need to recognise and capture serendipity and chance when it happens – this is associated with the willingness for risk-taking and inherent self-confidence and self-belief in the decision makers.

5. The opportunity to involve hybrid thinkers to help find hybrid solutions from non-traditional or non-conformist perspectives – being inspired by a fusion of ideas and an embrace of alchemy.

6. The willingness for exaptation (the process by which features acquire functions for which they were not originally adapted or selected) to take place.

A little bit of rebellion

The third question relates to the role and influence of religious nonconformity in the Cokers and the presence of dissenters – individuals and families who were prepared to be open to new and different ways of thinking about religion and, hence, society and its development.

It would appear that, despite the physical and religious dominance of the two Church of England churches, St. Michael and All Angels' in East Coker and St. Martin of Tours' in West Coker, there was a relatively large number of dissenters' chapels: St Roche Methodist Chapel in Moor Lane, East Coker; Baptist Chapel on Long Furlong Lane and a Methodist Chapel in Burton Lane in North Coker; a Wesleyan Chapel in West Coker and a Quaker Meeting Room in nearby Preston Plucknett.

In addition, Sir Jerome Murch (1835) writes about numerous nonconformist ministers arriving in the Yeovil area to establish chapels in the 18th century. Sims (2020) suggests that the nonconformist tradition was 'very important'. For the development of the rope, twine and sailcloth industries – and indeed for the development of Yeovil as a centre of glove-making. Sims points out that because their religion prevented them from holding usual positions in society in the military or in politics, the dissenters would concentrate their energies and activities on industry and commerce.

Whatever the reasons, the indisputable truth is that these small South Somerset villages created a phenomenon that captured the attention of great thinkers and key political influencers and inspired generations.

Glossary – A Language of its Own

Back basket – A washing basket used in bucking.

Barton and yarn barton – A yard or field for the drying of flax.

Bleaching – The process by which the colour is removed or made lighter by sunlight or chemicals.

Bolling or Breaking – The activity of breaking the stems of the plants to remove any woody parts, usually done by hand. Mechanisation was introduced locally in 1770.

Bolts – A bolt is a unit of measurement used as an industry standard for a variety of materials, typically those stored in a roll, from wood to canvas. The length is usually either 40 or 100 yards, but varies depending on the fabric being referred to – for example, a bolt of canvas is traditionally 39 yards.

Bucking and the bucking house – The term 'bucking' derives from the Middle English 'booker'. Buck was a lye, or liquor, in which fibres, yarn or cloth was steeped in order to bleach it. Bucking, therefore, was the process of bleaching and washing the fibres, yarn or cloth in lye. A bucking house was where the process took place.

Ell – A former measure of length used for textiles, equivalent to six handbreadths, locally variable but typically about 45 inches in England and 37 inches in Scotland. (The word means 'arm' and survives in the form of the modern English word 'elbow'.)

Flax and hemp dressing – The overall process, combining breaking, scutching and hackling (or heckling).

Hackling or combing – A hank of raw fibre was thrown into a hackle – a wooden board with upturned pins. The fibre was then drawn through the pins several times. The pins act like a comb to draw the fibres into one direction (to line the fibres) and to separate the short coarse fibres (tow) from the longer fine fibres.

Lead seals – Small leaden discs most commonly attached to textiles.

Lye – A strong alkaline liquor rich in potassium carbonate, leached from wood ashes and used especially in making soap and for washing.

Retting – This was conducted to remove fibres from the stems of the harvested plants through partial decomposition. Dew retting took place when plants were placed in thin layers on the grass and natural dew was the catalyst for fungal decay over 20 days. The process was reduced to 15 days when the plants were placed in a retting pond or in a slow-moving stream.

Ropejack – A machine used to impart twist into the ropes being made.

Scaling – The stripping of the hemp and flax from the remainder of the stem: this task was undertaken by women and children. The scaling was collected and burnt with the ash being spread on the fields.

Spinning – This process converts weak fibres into stronger yarn by giving them an overlap and a twist. The spinning was done in the home by women, with the help of children. Mechanical spinning was introduced locally in about 1807.

Scutching – The removal of woody parts of stems using a wooden baton attached to a short chain to beat the plants. This process was mechanised locally in 1803.

Twine walk – A narrow piece of land some 100m wide in which three or more yarn threads are twisted into twine.

Weaving – The method of fabric production in which two sets of yarn are interlaced at right angles to each other. The warp forms the vertical threads while the weft forms the horizontal ones.

The Sailmaker's Tools

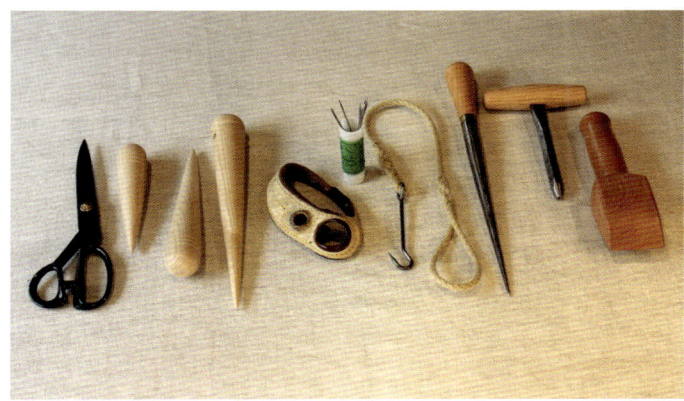

The following descriptions relate to the tools illustrated above, from left to right (courtesy of Mark Matthews, sailmaker).

Scissors – These are the cutting tools of a sailmaker. Scissors are precious to the individual sailmaker and are often personalised and marked for identification to their owner.

Wooden fids – These come in a range of sizes both in length and diameter tapering down towards a point. These are used for stretching worked rope cringles on the sails prior to inserting round metal thimble eyes which are fitted to protect the rope from wear and chafe at its attachment point. By wrapping thread around them they can also be used as a lever to tighten stitches on bolt ropes and hand sewn rings.

Palm – This is one of the key elements of a sailmaker's tool kit, worn around the working hand this offers protection and also has a dimpled metal insert to hold the back of the needle when pushing through heavy material.

Tube of sailmaker's needles – A range of different sizes for working on different weights and thickness of material.

Bench hook with lanyard – This is used as a third hand. When it is secured to the work bench by the lanyard it can then be hooked into the sail to help tension the fabric for ease of sewing.

Sail pricker/marline spike – These are metal fids and can be used to make small holes in material. They can also be used for rope work.

Heaving stitch mallet – Used to put additional tension into hand stitching when the thread is wrapped around the metal shaft and twisted tight.

Seam rubber – These are used for folding and rubbing down a seam or edge tabling of a sail. By adding pressure to the sail material it flattens it down and smooths it out prior to stitching.

Hand-sewing thread – The threads used on sails come in a range of materials, such as hemp, flax (linen), cotton and more recently polyester.

Beeswax – This is used to wax the thread for hand sewing, which enables two strands of thread to be twisted together and helps the thread to pass through the material being sewn. It can also be melted down to coat the strands of a rope that has been tapered down into a rat tail.

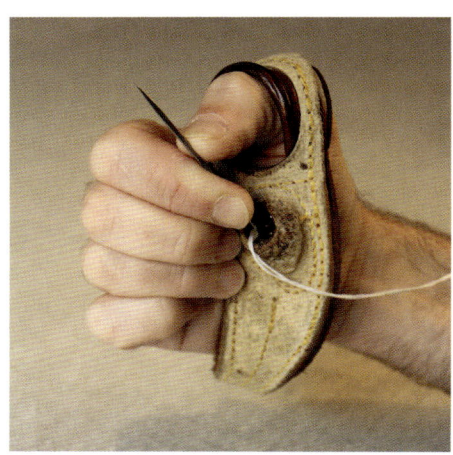

The Sailmaker's Tools

EAST COKER – *Esse Quam Videri*
Terry Stevens, March 2024

Time passes,
But doesn't diminish
Or erode
Identity.
The beginning.
Moor and marsh
Church and chapel.
Flax and hemp.
Rope, twine, and sail cloth.
Yarn.
Od, the crooked water.

Passing the gateway to many souls
Of the living and the dead.
Their ending.
One arrives.
The brow of the hill.
Church tower. Sentinel.
Its timeless clock framed by ancient yews and oaks.
I have returned.
From my journey. Like his.
Ancestors' past – and present.

Home is where you are.
In the moment
With loved ones.
Now.
The being never changes.
Things come and go.
He says houses rise and fall.
Live and die.
The being is consistent.
It just is. Always has been. Always will.
The future.
Now.
From the beginning to the end.
From the end to a new beginning.

Above: An ever-rattling green and white Bristol bus.

Above: Cottages, East Coker.

Bibliography, Sources of Information and Further Reading

A great deal of information relating to the business of running the Dawe's Twineworks is contained in four books:

'A Book of Composition and Costs of Various Twines', dated 1877.

'Petty Cash Book' – 1899 to 1914.

'Bought Ledger' – 1899 to 1914.

'Sales Ledger' – 1868 to 1874, which the Coker Rope and Sail charities has recently had restored.

Here are the details of sources and references to help if you want more information about the subject and would like to dive deeper into the subjects and characters in this book:

Ackroyd, P. (1985) *T. S. Eliot*. ABACUS. London.

Aitken, R. (2017) Conserving and Restoring Dawe's Twineworks. Bulletin No. 136 December. Somerset Industrial Archaeological Society. Taunton.

Barnes, G. (2006) 'Curiosity, Wonder, and William Dampier's Painted Prince'. *Journal for Early Modern Cultural Studies*, Vol. 6, No. 1 (Spring – Summer, 2006), pp. 31-50.

Batten, J (1894) *Batten's South Somerset Villages*, Whitby and Son, Yeovil.

Bennett, J.H. (1964) 'Cary Helyar, merchant, and planter of seventeenth-century Jamaica'. *William and Mary Quarterly* 21 (2).

Dunn, R.S. (2000) *Sugar & Slaves: The Rise of the Planter Class in English the West Indies 1624 – 1714*. Chapel Hill, North Carolina Press.

Eliot, V. (1969) *The Complete Poems and Plays of T. S. Eliot*. Book Club Associates. London.

Foot, D. (2010) *Footsteps from East Coker*. Fairfield Books. Bath.

Gardener, H. (1978) *The Composition of the Four Quartets*. Faber & Faber. London.

Gill, A. (1997) *The Devil's Mariner: William Dampier, Pirate and Explorer*. Michael Joseph. London.

Hackwell, B. A. (1957) *The Story of Our Village*. Reprinted by Swift Printing, East Coker, 1979.

Hann, A. & Dresser, M. (2013) *Slavery and the British Country House*. English Heritage. London.

Hore, P. (2004) *The Habit of Victory: The Story of the Royal Navy 1545 to 1945*. National Maritime Museum. London.

Little, B. (1969) *Portrait of Somerset*. Robert Hall. London.

Lloyd, C. (1966). *William Dampier*. Faber and Faber.

Mee, A (1968) *The King's England: Somerset*. Hodder & Stoughton. London.

Nathan, M. (1957) *The Annals of West Coker*. Cambridge University Press.

Norris, G. (Ed) (1994) *William Dampier: Buccaneer, Explorer*. The Folio Society. London.

Preston, D. & Preston, M. (2005) *A Pirate of Exquisite Mind*. Berkeley, California.

Prudden, H.C. (1993) *Geological Trails of Yeovil*. South Somerset District Council.

Purves, D.L. (ed.) (undated) *The Voyages of Sir Francis Drake and William Dampier: According to the Original Narrative*. W.P. Nimmo, Hay & Mitchell. Edinburgh.

Ramsden, D. (2015) *"Generally Fit to be Trusted": Social Networks and the Moral Economies of Sugar Plantations in Anglo-Jamaica*. Dalhousie University, Halifax, Nova Scotia. July.

Seymour-Junes, C. (2002) *Painted Shadow: The Life of Vivienne Eliot*. Constable. London.

Shepherd, A. (1997) *East Coker: A Village Album*. Coker Books.

Shorey, D. with Dodge, M. and Dodge, N. (2008) *The Book of West Coker: A Pictorial and Social History of a Somerset Village and its People*. Halsgrove.

Sims, R. (2009) *Rope, Net and Twine: The Bridport Textile Industry*. Dovecote Press.

Sims, R. (2015) *Sailcloth, Webbing and Shirts: The Crewkerne Textile Industry*. Studio 6 Publishing. Bridport.

Sims, R. (2018) *Coker Canvas: The Textile History of the Villages on the Somerset / Dorset Border*. Studio Six Publishing. Coker Rope and Sail Charities.

Sims, R. (2022) *West Coker in the Sailcloth Era*. CRS Monograph. Coker Rope and Sail Charities.

Sims, R. (2022) *Coker Canvas and the America's Cup Yachts*. CRS Monograph 8. Coker Rope and Sail CIO.

Stevens, T. (2020) *The Coker Canvas Cluster of Sail and Rope Making*. CRS Monograph No. 7. Coker Rope and Sail CIO.

Whipple. A.B. C. (1980) *The Seafarers: The Racing Yachts*. Time-Life Books. Amsterdam.

Wilkinson, C. (1929) *William Dampier*. John Lane, The Bodley Head Ltd. London.

Worthen, J. (2009) *T. S. Eliot: A Short Biography*. Haus Publishing. London.

Young, J. (2023) 'East Coker 0.2 after T. S. Eliot' second poem from *Four Quartets*. OSR> West Coker.

For information about Helyar family's estates in Jamaica see the University College London's Centre for the Study of the Legacies of British Slavery.

The Od Arts Festival is a project by OSR Projects Community Interest Company of West Coker founded in 2011 by Simon Lee Dicker and Chantelle Henocq: www.odartsfestival.co.uk; www.osrprojects.co.uk.

Pathe Film www.britishpathe.com/video/sailmakerstakecare.

www.trove.nla.gov.au

www.chroniclingamerica.loc.gov

Photo Credits

Lawrence Holden for permission to use photographs from the Dennis Chapman Archive.

Abigail Shepherd: pages 13 (bottom), 17, 58, 59 (top), 76 (right).

Alamy: pages 38-39, 88 (top left).

Alamy/2002 A.D Blake: pages 84-85.

Alan Stevens: page 69 (bottom).

Beken of Cowes: page 87.

Bob and Jean Richards/Abigail Shepherd: page 14.

Bob Osborn, Yeovil A-Z Virtual Museum: pages 49 (bottom), 52 (bottom), 53.

Bridport Museum: pages 42 (top right, bottom left and right), 47.

Bristol Weaving Mill: page 5.

Coker Rope and Sail Trust: pages 4 (bottom), 10, 13 (top), 20, 21, 48, 49 (top), 55, 69 (top), 70, 71, 72, 73.

David Chapman Archive: pages 59 (bottom), 83.

David Shorey: pages 23, 33, 63 (top), 64, 65, 101 (top left), 118.

Dominic Jones, Mary Rose Trust: page 79.

Emily Kerr: page 126.

Estate of John Masefield, Tomes Marime and National Gallery of Australia: page 96.

Gerry Smith: pages 63 (bottom left), 110.

Imperial War Museum: pages 41, 42 (top left, middle left and right), 43.

Job Vermeulen/America's Cup: pages 89, 90, 91.

John Snelling, Studio Elite: pages 18-19, 22, 25, 26, 28, 35, 36, 37, 61, 92, 93, 94, 100, 102 (top), 103, 105 (bottom), 108-109.

Louis Bartos: page 51.

Mark Matthews: pages 81, 82, 113, 123 (bottom).

Mary Evans Picture Library: pages 62, 80, 88 (right).

Mrs M Snell/Abigail Shepherd: page 45.

National Gallery of Australia, Canberra: page 97.

National Maritime Museum, Greenwich: page 50.

National Museum of the Royal Navy HMS *Victory* Archive V2009/562: pages 52 (top).

OSR Arts, West Coker: page 40.

Peter Gould: page 76 (left).

Portsmouth Historic Dockyard: page 121 (right).

Ralph Stevens/Gerry Smith: page 66.

Ratsey & Lapthorn: page 86.

Shaun Whitehouse: page 125.

Sidney Morning Times and Pacific Islands Monthly (National Library of Australia): page 67.

Simon Lee Dicker, OSR Arts: page 77.

Somerset Records Office / Abigail Shepherd: page 60.

Stevens Family Archive (originally National Library of Scotland): pages 12.

Stevens Family Archive: pages 4 (top), 24, 30, 101 (bottom), 107, 114, 115, 119, 122 (top right).

Suzanne Easton, Meandering Wild: pages 74, 75.

Terry Stevens: pages 8, 9, 11, 16, 31, 46, 57, 63 (bottom right), 102 (bottom), 104, 105 (top), 124 (bottom).

Wikimedia, public domain: page 78.

Above: Job Gould Twineworks outing to Cheddar and Weston-super-Mare, c. 1930.

Above: Drake's Webbing Factory, c.1950.

Grants, Supporters and Sponsors

Visit Somerset

Be enchanted, Visit Somerset. Welcome to Somerset, where lush countryside landscapes, historic charm, and captivating culture converge to offer an unforgettable experience. Nestled in the heart of South West England, Somerset beckons with its picturesque rolling hills, quaint villages, our traditional resorts, and meandering rivers. From the iconic Glastonbury Tor shrouded in myth and legend to the majestic Wells Cathedral boasting stunning architecture, Somerset is a treasure trove of heritage and history waiting to be explored. Indulge in the region's culinary delights with its renowned cider, delectable cheeses, and farm-fresh produce, or embark on outdoor adventures along the rugged coastline of Exmoor National Park. Whether you're seeking relaxation or adventure, Somerset promises a truly enchanting escape for every traveller. The story of Coker Canvas is a truly unique and authentic part of our heritage. A visit to an open day at the Dawe's Twineworks, the iconic attraction celebrating the rope and sail heritage of East, West and North Coker should be on your agenda. As Chair of Visit Somerset, the organisation responsible for prompting this wonderful part of Britain, I am delighted to support the work of the Coker Rope and Sail charities.

www.visitsomerset.co.uk

THE HEADLEY TRUST

The Headley Trust

The Headley Trust was founded in 1973 by Sir Timothy Sainsbury. It is one of 16 Sainsbury Family Charitable Trusts. The Headley Trust makes grants of around £5 million a year, largely in the arts and heritage field. Among its priority areas are museums and galleries, particularly those operating in the regions. The trustees particularly wish to support the professional development of curators, as well as the display, study and acquisition of British ceramics and the conservation of industrial, maritime and built heritage.

www.sfct.org.uk/the-headley-trust

Marc Fitch Fund

The Marc Fitch Fund is an educational charity established in 1956 by Marc Fitch (1908–1994). The Fund makes small grants towards the costs of publishing scholarly work in the fields of British and Irish national, regional and local history, archaeology, antiquarian studies, historical geography, the history of art and architecture, heraldry, genealogy and surname studies, archival research, artefact conservation and the broad fields of heritage, conservation and the historic environment.

ESE Capital

ESE Capital are committed to bringing about positive environmental and social change through each of the projects we bring to our platform. Born from a desire to revolutionise the housing market in the UK by providing genuinely affordable developments in areas of the country with high housing need, ESE Capital has matured to now include both residential and commercial projects in the UK and overseas. Our commitment to affordability is still at the heart of everything we do alongside our desire to provide a positive impact for local communities through increased infrastructure and the creation of new jobs.

ESE Capital is the consolidation of several long-standing businesses in the financial sector. This gives us the benefit of many decades of experience in providing high net worth and sophisticated investors with alternative investment opportunities at home and overseas.
We use this experience to ensure that the projects we bring to the ESE Capital platform are the very latest in socially conscious, ethical investments. As a company, we are committed to upholding the UN's 17 goals of sustainability through our use of sustainable materials, modern methods of construction and low carbon/carbon neutral targets for all of our projects.

We are delighted to be able to support this book in recognition of everything that is good about community heritage and sustainable development.

Greg Baker, Founder
www.ese-capital.com

Portsmouth Historic Dockyard

Embark on a voyage through time at Portsmouth Historic Dockyard, where maritime legends come to life! Step aboard HMS *Victory* and HMS *Warrior*, icons of naval history, and delve into their captivating stories. Explore fascinating artifacts and interactive displays, including a journey through the centuries at the Mary Rose Museum.

Portsmouth Historic Dockyard is proud to wish the Coker Rope and Sail Trust luck in their noble mission to preserve and interpret England's maritime heritage.

Above: The Mary Rose Museum, Portsmouth Historic Dockyard.

Grants and Sponsors

Somerset County Cricket Club

Somerset County Cricket Club is proud to represent the whole South West region and it's a unique strength of the Club having such a strong relationship with local towns and villages from across Somerset, Devon, Cornwall and Dorset. The Club has a strong tradition of local players coming through and playing professional cricket, whilst the Somerset Cricket Foundation does excellent work in linking in the recreational game under the SCCC umbrella.

We all wish the Coker Rope and Sail Trust success with its important work conserving all aspects of our local heritage.

Navigate

Navigate is a marketing consultancy that shapes success stories for tourism brands, large and small. With a compelling track record, over two decades, Navigate has crafted a specialism in growing visitors, income, and impact. Navigate's team of seasoned tourism marketing experts is dedicated to helping clients realise their ambitions. They have a strong connection to nature, and their mission of helping people explore and protect our world is at the heart of everything they do. They wish the Coker Rope and Sail Trust every success in its inspirational conservation work, which conserves and celebrates the rich and unique maritime connections.

Anian Consultancy

Anian is a destination marketing and place-making agency based in Cardiff. The word *anian* is Welsh for 'nature, characteristic, instinct...' – a profound sense of deep identity. I chose it because it is the name my father gave his own first company, and because – as he fully understood – the word captures the magical significance of protecting a place's innate spirit in order to give it new life. It captures the essence of sustainable place-making.

It was inspired by Welsh communities like St Davids in Pembrokeshire, where the spiritual is tangible today, and by my father's childhood home, East Coker, a village so rich in its sense of place, you can feel it glowing from the old stones around you.

But to respect a place's identity is not the same as allowing it to stand still. Thanks to T. S. Eliot, East Coker has become a village that signifies continuity through change, in a transient, modern world. The gentle babble of the roadside brook sounds the same today as it did 40 years ago, 100 years ago, 300 years ago.

The art of great place-making is to let that echo of the past drive the future; the threads of history weave into new and stronger sails. Anian is proud to sponsor this wonderful book – which aims to achieve just that.

Mari Stevens

Above: Lea-on, Moor Lane, East Coker.

MM Sail Solutions

Mark started as a sailmaker in 1990, aged 16, when he became an apprentice at Kemp Sails in Wareham. In over 30 years, his work has embraced many different types of sailmaking, from working on the Olympic dinghy sails in 2012 to those of large modern yachts. He has worked at many of the sail lofts on the South Coast, widening his experience, and in 2021 he became self-employed giving him the opportunity to pursue his own sailmaking interests.

Since 2018 he has been teaching sailmaking at the Boat Building Academy (BBA) in Lyme Regis and this has developed his interest in sailmaking for classic boats and for traditional techniques. Sailmaking is on the Heritage Crafts Red List of Endangered Crafts and Mark recently instigated a course in traditional sailmaking at the BBA.

Further research into traditional skills has led him to work with Ratsey & Lapthorn in Cowes and in 2023 Mark worked with Cowes Classic Boat Museum creating the sail for the 1872 built classic 'catboat' *Vigia*. It is this passion for the history of his subject that brought him to the Twineworks at West Coker to see their important work in the recreation of the renowned Coker Canvas.

Mark Matthews – Sailmaker

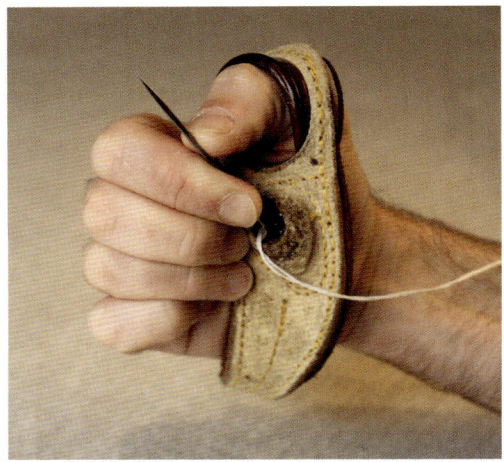

Boat Building Academy

At the Boat Building Academy in Lyme Regis, we are proud to offer professional training in boat building and furniture making in the heart of the UNESCO WHS of the Jurassic coast. People of all ages and backgrounds sign up for anything from 2 days to 40 weeks to pursue training for a new career, or for the sheer joy of learning new skills in a welcoming and inspirational environment. We are proud to keep the skills alive and as a charity offer bursaries to make the training accessible to all. We wish the Coker Rope and Sail Trust all the best in their endeavours.

Will Reed, Director and Co-chair of Trustees

www.boatbuildingacademy.com

Non Stevens and Mike Normansel and the boys

In fondness of a special place wrapped in happy memories of holidays at Lea-on, walks in Coker Wood, turnip rolling competitions at Ham Hill, and the family gatherings in the village pubs. So many stories, so many characters, so much history. We wish Coker Rope and Sail Trust every success in all its efforts to celebrate the heritage of Coker Canvas and look forward to more visits to the Dawe's Twineworks in the coming years.

Other sponsors

Coles Holdings

Ross Aitken

Ms. Yoga Gresnerova

and others who wish to remain anonymous.

John and Peter Gould

Twine manufacturing with the Goulds in West Coker. Our family have lived in West Coker and been involved with the rope and twine manufacturing industry for several hundred years. Job Gould, through his good fortune in receiving an inheritance from his step grandfather, Captain Herbert, was able to obtain ownership of the land that became the West of England Twine Works and he built the twine walks and office buildings which are now Grade II listed; rope and twine manufacturing has been carried on there by five subsequent generations. Originally, Job Gould bought the raw material, flax, grown locally and harvested by pulling from the ground, as the main strength of the fibres is at the base of the stem, and processed in a steam heated retting tank before taking it to the two hackling sheds, producing the fibre, enabling him to control the quality of the yarn; the fibre was spun into yarn by Pymore Mill at Bridport and brought back to the twine works to be made into the high quality twine for the upholstery trade. We can remember when all the old twine walks were used to twist and finish the cordage and twine, later there was also a parallel system of machine twisting and finishing which produced a similar product but not as good as the traditional walk process. We came to appreciate the skills and loyalty of everyone who worked at the factory without whom it could not have flourished. We give our best wishes to the Coker Rope and Sail Trust who are doing a great job in celebrating the history of manufacturing in our village.

Ratsey & Lapthorn

Ratsey & Lapthorn are the oldest sailmakers in the world and have been making sails in Cowes on the Isle of Wight since 1790. The skills in the loft have been handed down by generations and today are combined with the latest up-to-date design and manufacturing technology for the perfect blend of the old and the new. Our team consists of expert sail designers, sailors and sailmakers. We are knowledgeable and passionate and pride ourselves on providing a truly bespoke service. There is no boat too big or too small – from Optimist to Schooners, we welcome every boat we make sails for into the Ratsey family.

Ratsey & Lapthorn is synonymous with the America's Cup, the oldest sporting trophy in the world, first competed in 1851. R&L were the sailmakers of choice, providing sails for competing yachts from both sides of the Atlantic for over 100 years.

The company has opened a dedicated service, repair and manufacturing loft in Marina Vela, Barcelona in 2019 to service our Mediterranean-based customers and to mitigate concerns and logistical challenges as a result of Brexit. Our Ratsey team in Barcelona is led by Gonzalo Romagosa and are on hand in the host city of the 37th America's Cup ready to manage sailors' needs for new sails, covers, design, repairs, rigging or advice.

#ratseyandlapthorn

www.ratseyandlapthorn.com

Lanes Hotel

Lanes... where traditional Ham stone meets cool, seamless glass. This former rectory has been sympathetically renovated to give a modern, contemporary feel to the bar and restaurant areas whilst retaining the elegance and individuality of the 10 spacious bedrooms above. A modern extension building echoes these design features and houses a further 16 well-appointed rooms, a gym and spa. There are also 4 fully self-contained 1 and 2-bedroom apartments. Our popular and lively bar features a range of local Somerset ciders, in-house crafted and aged cocktails as well as an extensive beer and spirit choice. Our restaurant, Lanes Brasserie, serves local produce in a light and uncomplicated brasserie style, complemented by a well-chosen wine list. Our menu rotates through the seasons to bring you the freshest ingredients and outstanding quality.

This 18th-century Grade II listed former Rectory has always been at the heart of village life in West Coker – serving the community in many different ways and being the former home of the Gould family (sailcloth makers). In the 1960s it was bought by the Hannam family and converted into the Somerset Hotel and, later, the Coker Motel. We are delighted to support the work of the Coker Rope and Sail Trust and congratulate all involved in sharing the remarkable story of Coker Canvas with the wider world.

Above: Lanes Hotel, West Coker (former 18th-century rectory).

Emily Kerr Photography

Enthralled by the interplay of vibrancy and shapes, Emily infuses her artistic essence into a vibrant tapestry of bold and colourful compositions. Emily's fascination with reflections is unmistakable – a fascination that unveils the beauty of duality, allowing viewers to witness the convergence of two scenes into a singular, captivating narrative. It's through the magic of light and water that she orchestrates moments of enchantment, where the fluidity of movement intertwines with the brilliance of illumination. Emily's photography serves as a testament to her unyielding passion for storytelling through imagery. With every frame, she invites her audience to embark on a journey of discovery, where the ordinary transcends into the extraordinary.

Based in Barcelona, Emily is delighted to support the work of the Coker Rope and Sail Trust and its celebration of the deep heritage story associated with the America's Cup.

Above: *Afloat in Reflections.*

Museu Marítim de Barcelona

In the field of maritime history and culture, nothing is too small, too far, or too local. The ocean is one, unique, and men and women who live in it or for it are quite similar anywhere. They have the same jobs, sorrows, fears and expectations.

On the occasion of the 37th America's Cup in Barcelona we have had the chance to discover the Coker Rope and Sail Trust at the Dawe's Twineworks, an extraordinary story which some time ago was absolutely unknown to us.

The experiences in the recovery of old maritime heritage sites are always welcomed. If this challenge includes keeping traditional skills alive, it gets better and better. One of the most outstanding values in this particular case is that the factory, the technology and the skills are only a part of this recovery, while the social dimension of the trade is still to be vindicated. The stories of children, women and men should have to coexist among the factory walls, roofs and engines. This maritime community lived and suffered together to provide seamen with the best ropes and sails. Their stories should be heard.

Any opportunity to collaborate with other organisations of maritime heritage sites is a gift for the Museu Marítim de Barcelona, no matter their location or size. So, here we are, sending our greetings to sea lovers and seamanship, wherever they are.

We are two different organisations, in more than one sense, but, as it could be expected, we have found a brotherliness that explains this short text.

www.mmb.cat

Dr Enric Garcia-Domingo
Director, Museu Marítim de Barcelona

NJOY Catalonia

NJOY Catalonia is a travel agency that provides a wide range of immersive experiences, getaways, and tours in Catalonia, focusing on enhancing travelers' experiences through culture, gastronomy, nature heritage, and traditional richness by immersive and multi-sectoral travel proposals. They are committed to sustainability and they cater to eco-friendly travelers to contribute to the preservation of the places they visit. Their eco-commitment: A traveler, a tree.

We would like to wish the Coker Rope and Sail Trust every success with all their projects.

www.njoycatalonia.com

Terry Stevens

Terry Stevens is an international tourism consultant. A love of travel and interest in landscape and heritage was inspired by Hugh Prudden, a teacher of geography at Yeovil Grammar School, and fuelled by fellow East Coker 'villagers' William Dampier (1651–1715), explorer, hydrographer and buccaneer, and the great 20th-century poet T. S. Eliot (both of whom are central characters in this book).

Terry studied Geography at Swansea University in 1970. After gaining his MSc in Recreational Land Management at Reading University, he developed tourism projects on private Estates in Hampshire and Dorset before returning to Wales to work for the Wales Tourist Board, he received his doctorate from the University of Wales and Bangor and has lived in Wales ever since with his wife Catrin. Their two daughters, Mari and Non, live with their families in Cardiff.

His early career was spent on all aspects of leisure, tourism and heritage for the Pembrokeshire Coast National Park, West Glamorgan County Council and Cadw: the Welsh Historic Monuments before being appointed as Professor and Dean of Tourism, Leisure, and Health Care Management in Swansea. During this time, he co-founded the East Coker Society and was also Director of the Safety in Leisure Research Unit, the Director of the UK Stadia and Arena Management Unit, a Member of the Countryside Commission of Wales, Vice-Chair (Strategy) of the Wales Tourist Board and Advisor to the European Centre for Cultural Tourism in Barcelona.

In 1986 he established his multi-award-winning international tourism consultancy, Stevens & Associates, and has now worked on destination development and management in almost 60 countries around the world. Clients have included the UN World Tourism Organisation, the World Bank, the European Bank of Re-construction as well as national and regional tourist boards and many private sector clients.

Terry is a Trustee of the Coker Rope and Sail Trust and of the Mary Rose Trust. He is a lifelong fan of the famous Glovers (Yeovil Town Football Club) for whom he played in the youth team and occasionally the reserves.

Among his numerous awards are the American EXPRESS Future of Tourism Award, three times winner of the LUXLife award for being the Best Destination Development Expert, and the Slovenian Government's Award for contribution to tourism. Terry has over 350 published works including *Landscape Wales*, *Wish You Were Here* and *Wish You Were Here Europe* (all published by Graffeg).

www.tourism-futures.com